AA002331

V International Symposium on Coherent Optical Radiation of Semiconductor Compounds and Structures 2015

Journal of Physics: Conference Series Volume 740

Moscow, Russia
23 – 26 November 2015

Editors:

O.N. Krokhin   **A.A. Fronya**
I.N. Zavestovskaya   **O.K. Bankova**
A.A. Ananskaya

ISBN: 978-1-5108-2915-2
ISSN: 1742-6588

Printed from e-media with permission by:

Curran Associates, Inc.
57 Morehouse Lane
Red Hook, NY 12571

Some format issues inherent in the e-media version may also appear in this print version.

Copyright© (2015) by the Institute of Physics
All rights reserved. The material featured in this book is subject to
IOP copyright protection, unless otherwise indicated.

Printed by Curran Associates, Inc. (2016)

For permission requests, please contact the Institute of Physics
at the address below.

Institute of Physics
Dirac House, Temple Back
Bristol BS1 6BE UK

Phone: 44 1 17 929 7481
Fax: 44 1 17 920 0979

techtracking@iop.org

Additional copies of this publication are available from:

Curran Associates, Inc.
57 Morehouse Lane
Red Hook, NY 12571 USA
Phone: 845-758-0400
Fax: 845-758-2633
Email: curran@proceedings.com
Web: www.proceedings.com

V International Symposium on Coherent Optical Radiation of Semiconductor Compounds and Structures 2015

Journal of Physics: Conference Series Volume 740

Moscow, Russia
23 – 26 November 2015

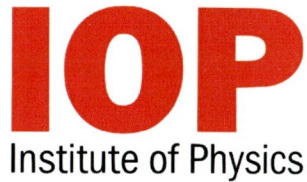

Table of contents

Volume 740

V International Symposium on Coherent Optical Radiation of Semiconductor Compounds and Structures

23–26 November 2015, Moscow, Russia

Accepted papers received: 1 August 2016
Published online: 1 September 2016

Preface

011001
OPEN ACCESS
Vth International Symposium on Coherent Optical Radiation of Semiconductor Compounds and Structures(Moscow - Zvenigorod, 23-26 November, 2015)

O N Krokhin, I N Zavestovskaya, A A Ananskaya, A A Fronya and O K Bankova

011002
OPEN ACCESS
ORGANIZERS of the V International Symposium on Coherent Optical Radiation of Semiconductor Compounds and Structures Moscow - Zvenigorod, November 23-26, 2015

011003
OPEN ACCESS
PROGRAM AND ORGANIZING COMITTEES Of the V International Symposium on Coherent Optical Radiation of Semiconductor Compounds and Structures Moscow - Zvenigorod, November 23-26, 2015

011004
OPEN ACCESS
Peer review statement

Papers

012001
OPEN ACCESS
Synchronization of semiconductor laser arrays with 2D Bragg structures

V R Baryshev and N S Ginzburg.....1

012002
OPEN ACCESS
Theory of operating characteristics of a semiconductor quantum well laser: Inclusion of global electroneutrality in the structure

Z N Sokolova, N A Pikhtin, I S Tarasov and L V Asryan.....7

012003
OPEN ACCESS
Diode lasers with front surface high-order distributed Bragg reflector

V V Zolotarev, A Yu Leshko, Z N Sokolova, Ya V Lubyanskiy, N A Pikhtin, D N Nikolaev, V V Shamakhov and I S Tarasov.....14

012004
OPEN ACCESS
The peculiarities of spectra in high power 970 laser diodes

V V Bezotosnyi, O N Krokhin, V A Oleshchenko, V F Pevtsov, Yu M Popov and E A Cheshev.....20

012005
OPEN ACCESS
The study of the laser characteristics based on solid solution $Pb_{1-x}Sn_xSe$ (x ~ 0.07) emitting at spectral range of 16 µkm

K R Umbetalieva, K V Marem'yanin, V I Gavrilenko, I I Zasavitskij, R R Bitskiy and E A Komochkina.....25

012006
OPEN ACCESS
Middle infrared Fe^{2+}:ZnS, Fe^{2+}:ZnSe and Cr^{2+}:CdSe lasers: new results

V I Kozlovsky, Y V Korostelin, Y P Podmar'kov, Y K Skasyrsky and M P Frolov.....32

012007
OPEN ACCESS
Empirical mode with a variable spatial-temporal structure and the dynamics of superradiant lasers

E R Kocharovskaya, A S Gavrilov, V V Kocharovsky, E M Loskutov, D N Mukhin1 A M Feigin and Vl V Kocharovsky.....38

012008
OPEN ACCESS
InAsP/AlGaInP/GaAs QD laser operating at ~770 nm

A B Krysa, J S Roberts, J Devenson, R Beanland, I Karomi, S Shutts and P M Smowton.....51

012009
OPEN ACCESS
High quality $Y_3Al_5O_{12}$ doped transparent ceramics for laser applications, role of sintering additives

A A Kaminskii, V V Balashov, E A Cheshev, Yu L Kopylov, A L Koromyslov, O N Krokhin, V B Kravchenko, K V Lopukhin, V V Shemet and I M Tupitsyn.....59

012010
OPEN ACCESS
8 nm nanodiamonds as markers for 2 photon excited luminescent microscopy

A Kharin, A Rogov, A Geloen, V Lysenko and L Bonacina.....68

012011
OPEN ACCESS
Permittivity and permeability of semi-infinite metamaterial

O V Porvatkina, A A Tishchenko and M N Strikhanov.....74

012012
OPEN ACCESS
Generalized Clausius-Mossotti relation for semi-infinite artificial periodic structure

M N Anokhin, A A Tishchenko and M N Strikhanov.....79

012013
OPEN ACCESS
Aberration influenced generation of rotating two-lobe light fields

S P Kotova, N N Losevsky, D V Prokopova, S A Samagin, V G Volostnikov and E N Vorontsov.....84

012014
OPEN ACCESS
Radioprotective Action of Low-Intensity Light into the Red Absorption Band of Endogenous Molecular Oxygen

A V Ivanov, A A Mashalov and S D Zakharov.....93

012015
OPEN ACCESS
Transmission of large amounts of scientific data using laser technology

E A Isaev and P A Tarasov.....96

012016
OPEN ACCESS
Funds support the construction of control systems design

V Vlasov and A Tolokonsky.....102

012017
OPEN ACCESS
Polymer-Free Carbon Nanotubes Saturable Absorbers for Nanosecond Pulse Generation

A V Zasedatelev, V I Krasovskii, O Reynaud, Yu G Gladush, D S Kopylova, E A Komochkina, E I Kauppinen and A G Nasibulin.....106

Vth International Symposium on coherent optical radiation of semiconductor compounds and structures
(Moscow – Zvenigorod, 23-26 November, 2015)

O N Krokhin[1,2], I N Zavestovskaya[1,2], A A Ananskaya[1,2], A A Fronya[1,2] and O K Bankova[1]

[1]National Research Nuclear University MEPhI, 115409, Moscow , Russia
[2]P.N. Lebedev Physical Institute of the Russian Academy of Sciences, 119991, Moscow, Russia

E-mail: INZavestovskaya@mephi.ru

1. Introduction

The Vth International Symposium on coherent optical radiation of semiconductor compounds and structures is recognized as an important scientific research platform with participation of the leading academic institutions working on modern directions of fundamental research in coherent optical radiation of semiconductor compounds and structures.

The major topics of the Symposium are: semiconductor lasers on the basis of heterostructures; semiconductor lasers with optical and electronic dumping; unipolar semiconductor lasers; current trends in coherent optical sources; physics of fracture and degradation of radiation semiconductor structure etc.

Currently, semiconductor lasers are most in demand in quantum electronics [1, 2]. To achieve more success in quantum electronics further improvements in semiconductor laser specifications are needed. These improvements are to be based on new approach and modernization of techniques which require a better understanding of physical processes.

Sustainable development of the domestic laser and optoelectronics engineering largely depends on new approaches and the scale of research in this area [3]. The closest attention should be given to the research and development of semiconductor injection lasers operating in the blue and near-end ultraviolet band spectrum. More efficient light sources can be provided with the help of lasers rather than light-emitting diodes because laser efficiency is higher than that of light–emitting diodes.

Cooperation of research groups of academic institutions with Russian manufacturing enterprises specializing in semiconductor lasers and optoelectronic systems is crucial to the sustainable innovative development of these enterprises and the branch on the whole.

The first Symposium on semiconductor coherent optical radiation compounds and structures was held in 2007 in Zvenigorod [4]. Its participants adopted a decision to hold Symposium on a regular basis once every two years. The IId, IIId and IVth Symposiums were held in 2009, 2011 and 2013 respectively. To attract more participants and in particular young scientists, post-graduates and senior students it was decided to hold one Symposium session at P.N. Lebedev Physical Institute and two sessions in Zvenigorod.

The Vth International Symposium on semiconductor coherent optical radiation compounds and structures was held on 23-26 of November in 2015 in Moscow at P.N. Lebedev Physical Institute,

Content from this work may be used under the terms of the Creative Commons Attribution 3.0 licence. Any further distribution of this work must maintain attribution to the author(s) and the title of the work, journal citation and DOI.

Published under licence by IOP Publishing Ltd

National Research Nuclear University MEPhI and on the premises on the recreation house "Zvenigorodsky" in Zvenigorod, Moscow region. The event was organized by P.N. Lebedev Physical Institute, National Research Nuclear University MEPhI, Physics Department of the Russian Academy of Sciences, Russian Foundation for Basic Research, the journal "Quantum Electronics".

Scheduled sessions of the Symposium helped to reveal research results to the scientific community for a broader discussion, define the possible perspective trends of research, set essential tasks to research groups.

2. Participants

The Vth Symposium was attended by 256 participants, half of which were young scientists, post-graduate students and senior students. A series of 24-26 November sessions involved a plenary sitting, 6 scientific report sessions and an exhibition-stand session. In the framework of the Symposium a round-table discussion was organized to implement the Physics Department of RAS Program: "Fundamentals and experimental implementation of perspective semiconductor lasers for the benefit of industry and technology" (Research results for the years 2014-2015).

Thanks to a wide range of participants (scientists from Great Britain, Belarus, Moscow and other Russian cities: Nizhny Novgorod, Samara, Vladivostok, St. Petersburg) the results of the Symposium were revealed to the scientific community for a broader discussion, new essential tasks were set in the research of fundamentals of coherent optical radiation of semiconductor structures. 20 Russian organizations including academic institutions (Institutes and Universities) and manufacturing enterprises took park in the Symposium. Among other participants there were B.I. Stepanov Physics Institute of the National Academy of Sciences of Belarus; The Institute of Optical Sensor Systems DLR, Berlin, Germany; University of Sheffield; S1 3JD (Great Britain); School of Physics, Peking University, Beijing (China);Virginia Polytechnic Institute and State University (USA).

3. Agenda of the Symposium

The Vth Symposium major topics
1. The use of semiconductor lasers and solid-state lasers with diode pumping
2. Semiconductor lasers with optical and electronic pumping
3. Semiconductor lasers of THz band
4. Research of optical effects in semiconductors.

The majority of research projects focused on the development semiconductor laser technology and interaction of laser radiation with the matter.

Let's take a closer look at the most interesting studies done on each of the subject matters.

3.1. The use of semiconductor lasers and solid-state lasers with diode pumping

V.A. Orlovich (B.I. Stepanov Physics Institute of the National Academy of Sciences of Belarus) presented the results of his latest research of mini and microchip-lasers with dilatational (direct-axis) diode pumping and internally phase modulated laser non-linear optical transformation. His report touched upon the possibility of control over timing parameters of radiation transformation. He demonstrated generation of radiation of average power output up to 10 W with basic harmonic frequency and up to 3W with double frequency in the Nd:KGW laser; generation of nanosecond pulses of 3 Stokes lines with wave length band 1.2-1.6 micron and pulses of the 1st Stokes line of 50ps. In scientific paper "Generation threshold and synchronization of traversal modes of Ramanovsky radiation component of $Nd:YVO_4$ laser with passive shatter and lateral diode pumping" V.V. Bezotosny, M.V. Gorbunkov and others studied the effects of simultaneous synchronization of lateral and traversal modes. Generation threshold and pulse duration were studied in the resonator modes degeneracy areas in Nd:YFL laser with Cr4+:YAG passive shuter.

The report of A.B. Krysa from Sheffield University was devoted to the research to the possibility of expanding spectral band of InP quantum dots (QD) lasers towards the longwave due to adding arsenic during MOC-hybrid epitaxy in InP QD (arsenic concentration in crystal subgrid of the Vth group of

InAsP being 25%). Electroluminescence of InAsP QD laser structure demonstrated a longwave shift of QD radiation to 775nm (in comparison with the studied test sample with binary InP QD at 716nm wavelength).

A.A. Marmalyk report (M.F. Stelmakh Research Institute "Polus") presented a perspective strategy of developing powerful emitters through increasing electron confinement in quantum wells in the active area of laser heterostructure in case small thickness of waveguide layers remains constant.

The scientific paper "Laser diode arrays and matrices in lateral and traversal pumping systems of solid-state lasers" experimentally proved efficiency of single-frequency coaxial-cavity semiconductor lasers on the basis of Fiber Bragg Gratings in powerful narrow-band solid-state lasers for wind lidars. Also, it provided samples of diode pumping lasers designed by the Physics Institute of the National Academy of Sciences of Belarus to be used in distance and spectroscopic lidar measurement techniques.

3.2. Semiconductor lasers with optical and electronic pumping, UV and IR band semiconductor lasers
E.A. Stepanov research paper presented experimental results of digital modelling of ultrashort laser impulses compression at the wave length 4-8 mcm in gallium arsenide.

A high degree of time impulse compression was reported to have been achieved (less than two field cycles) in the studied wavelength band in the areas of normal and abnormal dispersion of gallium arsenide group velocity modulation. It should be noted that limiting duration value of the formed impulses is less than one field cycle (less than 20fs at the wavelength 6.5 μm).

The research paper "The source of UV mid-band with electron beam pumping on the basis on multi-layer heterostructure with quantum wells AlGaN" (V.I. Kozlovsky) demonstrated the use of heterostructures with 40 quantum wells spread along the thickness of 1.3 μm matched with the depth of the excitatory area. The pulse mode of electron beam pumping resulted in the power output of 160 mW noncoherent radiation with the efficiency factor of 0.7 % at the maximum wavelength of 285nm and 120mW with the efficiency factor of 0.5% at the max wavelength of 260 nm. In continuous pumping mode the power output reached 40 mW and 20mW at the wavelengths 285 and 260 nm respectively. To increase the power output of radiation the surface of the structure was divided into cells and coated with Al reflector. Ways of increasing UV source efficiency by electron beam pumping and design of lasers using similar structures are discussed in the article.

V.I. Kozlovsky presented also the report about Fe^{2+}:ZnSe и Fe^{2+}:ZnS lasers parameters at low temperatures with free-running Er:YAG laser pumping.

Also together with other researches new room-temperature results were received on crystals grown at the P.N. Lebedev Physical Institute.

Cr^{2+}:CdSe impulse laser (200ns, 10kHz, 2.92 μm) was designed with average capacity 0.35W with laser pumping radiation on Tm^{3+}:Lu_2O_3 ceramics laser

3.3. THz band semiconductor lasers
In report "THz frequency band lasers on donors in silicon" V.N. Shastin provides a brief overview of the recent research results related to the research of stimulated radiation of optically excited neutral donors of the 5th group in unaxially distorted crystal silicon. Also, it's considered the possibility of expanding frequency to 9.6 THz with deeper Mg donors.

A.A. Andronov introduced new intraband semiconductor GaAs (150A, well) - GaAlAs lasers on the basis of simple superlattice and 19A Wannier-Stark lasers, with 12% of share of aluminum, barrier). The gain mechanism in lasers is based on the population inversion between the basic Wannier-Stark level in superlattice wells and poorly populated top Wannier-Stark level in wells with 2,3,4 periods down the input potential. A.A. Andronov demonstrated that negative conductivity responsible for radiation remains at 300 K.

THz radiation didn't appear due to low gain in contrast to the cavity loss. In case of optimization such superlattices could become rivals to cascade lasers as sources of radiation from GHz to THz band and higher.

3.4. Optical effects in semiconductors. Laser materials of the future

In report "Photoinduced low threshold optical nonlinearity of dielectric nanosystems" V.P. Dzyba suggested a theoretical model of mechanism of optical nonlinearity dielectric nanocomposite medium in weak fields of laser radiation. This model explains experimentally observed singularities.

The main optical parameters of the studied ceramics such as "in-line" absorbtion/transmission spectrum; luminescence spectrum; ceramic grains size; the remaining pit holes volume and their volume distribution were investigated in the scientific paper "Optical properties of doped transparent ceramics Y_2O_3 и $Y3Al5O12$". Phenomenological model showing the impact of sintering additives on kinetics of the process of sintering was introduced. The best ceramics samples showed optical transmission equal to theoretical value of a monocrystal.

S.D. Zakharov presented preliminary results of clinical research on optical method to protect women suffering from breast cancer and chemotherapy side-effects. Low intensity diode lasers are used as sources of radiation. This method is based on light-oxygen effect which proved to be effective at all levels of biological substances from protein solution to a human substance. Light-oxygen therapy is effective at first signs of irradiation injury, also it helps to ease side-effects and prevent treatment gaps.

4. Russian scientists' contribution

The major research laboratories and research groups conducting studies in this area both in Russia and abroad participated in the Symposium. Leading foreign and Russian scientists took part in the plenary sessions and delivered reports at the Symposium sessions. O.N. Krokhin and Yu.M. Popov are well-known as designers of the first semiconductor lasers. They have developed courses in "Photonics" and "Semiconductor quantum generators" at National Research Nuclear University MEPhI for the students of related branches of studies. Academician O.N. Krokhin is the coordinator of the Branch of Physical Sciences of Russian Academy of Sciences Programs "Fundamental Physics and Production technique of semiconductor lasers as basic elements of Photonics and Quantum Electronics". Also he is the scientific adviser at Master's Degree Program Institute and N.G. Basov Higher School of Physicists, P.N. Lebedev Physical Institute and National Research Nuclear University MEPhI – main suppliers of specialists for industry and scientific laboratories. Yu.M. Popov is the head of the semiconductor lasers research group - at P.N. Lebedev Physical Institute.

G.T. Mikaelyan is the head of Research and Production Enterprise "Inject" (Saratov). The enterprise manufactures powerful laser diodes and diode lines.

A.A. Andronov - Doctor of Physical and Mathematical Sciences, corresponding member of Russian Academy of Sciences, member of the American Physics Society and Deputy Director of Institute for Physics of Microstructure of Russian Academy of Sciences (IPM RUS -Nizhny Novgorod) since 1993. He has published more than 100 scientific papers, designed laser and masers on "hot holes". He is also a recognized specialist in the arear of radio-physics and solid-state physics.

V.A. Orlovich - the National Academy of Sciences of Belarus Academician- is well-known as a leading specialist in the area of Raman scattering. He is the author of 9 inventions and 500 scientific publications. His research is devoted to nonlinear optics of laser physics, spectroscopy, and photobiology.

Also the leading school Ioffe Institute, P.N. Lebedev Physical Institute, Institute of Applied Physics of the Russian Academy of Sciences and others organizations were represented on the Symposium.

In conclusion of the Vth Symposium it was noted that the reports are of great interest, both from a scientific point of view and from the point of view of practical applications. Part of the presented works has a direct access to the practical application of the technology through the introduction in the production of semiconductor lasers and emitters based on them.

5. Human resourcing for the branch

Most of the research projects were supported by grants of Russian Foundation for Basic Research and Federal Target Program – "Scientific and Academics training of human resources for Innovative

Russia". The final proceeding of the Vth Symposium pointed out an active participation and interest of young scientists and specialists of the branch.

Rapid growth of semiconductor engineering in Russia and in the whole world calls for the development of new semiconductor lasers and wider field of their application. All the above stated requires highly-qualified human resources for the branch including timely training of young specialist by educational institutions. Currently National Research Nuclear University MEPhI in particular runs a Master's degree program "Physics and technology of semiconductor lasers" aimed at training specialists for laser technology, semiconductor quantum electronics, interaction of radiation with matter, photonics, semiconductor laser engineering and their application in different areas. The Program is to be delivered in English for foreign students.

The Master's Degree Program is implemented by the National Research Nuclear University MEPhI academic department № 88 "Semiconductor Quantum Electronics". P.N. Lebedev Physical Institute of Russian Academy of Sciences (Moscow) together with Research and Production Enterprise "Inject" (Saratov) is seen as the basic platform for the Program "Physics and technology of semiconductor lasers". The Program is headed by O.N. Krokhin - Professor, Doctor of Physical and Mathematical Sciences, Academician of Russian Academy of Sciences and head of the department.

Short-term on-the-job trainings have become a regular practice both at foreign and Russian enterprises. In the framework of the Symposium field training was arranged for the students of foreign, regional and Moscow universities. Master Degree Program students of the Russian Academy of Sciences academic department № 88 "Semiconductor Quantum Electronics" took part in the field training.

6. Conclusion

The Vth International Symposium on semiconductor coherent optical radiation compounds and structures was held on 23-26 of November in 2015 in Moscow (24 November) and on the premises on the recreation house "Zvenigorodsky" in Zvenigorod, Moscow region. The scientific reports presented at the Symposium have great scientific value. The program of the Symposium covered all the modern and recent trends in fundamental research in the area of coherent optical radiation of semiconductor compounds and structures. 256 participants took part in the Symposium among them scientists and researchers from many Russian cities and towns: Moscow, St Petersburg, Samara, Vladivostok, Nyzhny Novgorod and others, also from the CIS countries and from abroad. Half of the participants were young scientists, post- graduate students and senior students. A series of 24-26 November sessions involved 4 plenary sittings, a scientific report session (18 reports) and an exhibition-stand session (30), a round-table discussion. All the reports were delivered at a very high level and were followed by a constructive discussion sessions. At the final session of the Symposium on 26 November 2016 the organizers and participants admitted a very high scientific value and organization of the event. The program of the Symposium was fully completed. The Symposium adopted a decision to hold similar symposiums on a regular basis – once every two years. The next one is to be held in 2017.

Acknowledgements

We want to express our sincere gratitude to the organizers of the Vth Symposium in particular to its chairman - Academician O.N. Krokhin for the high level of organization of the event, for the support to "Russian Foundation for Basic Research", Competitiveness Program of National Research Nuclear University MEPhI, RAS Physics Department, RAS Physics Department Program "Fundamentals and experimental implementation of perspective semiconductor lasers for the benefits of industry and technology". Our sincere gratitude is offered to the LPI RAS Publishing House.

References
[1]   Popov Yu M 2011 The early history of the injection laser *Phys. Usp.* **54** 96–100
[2]   Alferov Zh I, Kromer G, Kilby J S 2002 Nobel lectures in physics – 2000 *Phys. Usp.* **172** 1067

[3] Technology Platform Photonics http://photonica.cislaser.com/
[4] Zavestovskaya I N, Krokhin O N, Popov Yu M, Semenov A S 2008 Symposium on the coherent optical radiation of semiconductor compounds and structures (Zvenigorod, 27—29 November 2007) *Quantum Electronics* **38** 294–7

ORGANIZERS
of the V International Symposium on Coherent Optical Radiation of Semiconductor Compounds and Structures
Moscow – Zvenigorod, November 23-26, 2015

- Division of Physical Sciences of the Russian Academy of Sciences
- P.N. Lebedev Physical Institute of the Russian Academy of Sciences
- National Research Nuclear University MEPhI
- Russian Foundation for Basic Research
- Quantum Electronics journal

PROGRAM AND ORGANIZING COMITTEES
Of the V International Symposium on Coherent Optical Radiation of Semiconductor Compounds and Structures
Moscow – Zvenigorod, November 23-26, 2015

Program Committee Chair
Academician Oleg N. Krokhin (P.N. Lebedev Physical Institute, MEPhI)
Program Committee Vice-Chair
Victor A. Zayats (Division of Physical Sciences of the Russian Academy of Sciences)

Program Committee

V. Kocharovsky (Institute of Applied Physics of the Russian Academy of Sciences)
S. Kotova (Samara Branch of the P.N. Lebedev Physical Institute)
Yu. Kulchin (Far East Division of the Russian Academy of Sciences)
V. Makarov (Moscow State University)
Yu. Popov (P.N. Lebedev Physical Institute, MEPhI)
I. Tarasov (Ioffe Institute)
V. Shastin (Institute for Physics of Microstructures of the Russian Academy of Sciences)
A. Krysa (UK National Centre for III-V Technologies, Sheffield)
V. Orlovitch (B.I. Stepanov Institute of Physics of the National Academy of Sciences of Belarus)
A. Kanavin (P.N. Lebedev Physical Institute, MEPhI)
N. Kargin (P.N. Lebedev Physical Institute, MEPhI)
E. Cheshev (P.N. Lebedev Physical Institute, MEPhI)
V. Bezotosnyi (P.N. Lebedev Physical Institute)
A. Parfenov (Physical Optics Corporation,Torrance, USA)
A. Strigazzi (Politecnico di Torino, Italy)
Ž. Pavićević (University of Montenegro)
M. Zdorovets (L.N. Gumilyov Eurasian National University, Republic of Kazakhstan)
A. Kabashin (Aix-Marseille University, France)

Organizing Committee Chair
Academician Oleg N. Krokhin (P.N. Lebedev Physical Institute, MEPhI)
Organizing Committee Vice-Chair
Irina N. Zavestovskaya (P.N. Lebedev Physical Institute, MEPhI)
Organizing Committee Secretary
Anna A. Shestukhina (P.N. Lebedev Physical Institute, MEPhI)

Organizing Committee

E. Manykin (MEPhI)

G. Mikaelyan ("Inject" Scientific-Production Association, Saratov)

A. Semenov (Quantum Electronics Journal)

A. Starodub (P.N. Lebedev Physical Institute)

V. Kozlovsky (P.N. Lebedev Physical Institute, MEPhI)

A. Tarasova (P.N. Lebedev Physical Institute)

V. Brevi (P.N. Lebedev Physical Institute)

A. Ananskaya (MEPhI, P.N. Lebedev Physical Institute)

I. Tupitsyn (MEPhI)

A. Fronya (P.N. Lebedev Physical Institute, MEPhI)

S. Leonova P.N. Lebedev Physical Institute)

O. Rodionova (MEPhI)

N. Karpov (MEPhI)

Peer review statement

All papers published in this volume of *Journal of Physics: Conference Series* have been peer reviewed through processes administered by the proceedings Editors. Reviews were conducted by expert referees to the professional and scientific standards expected of a proceedings journal published by IOP Publishing.

Content from this work may be used under the terms of the Creative Commons Attribution 3.0 licence. Any further distribution of this work must maintain attribution to the author(s) and the title of the work, journal citation and DOI.
Published under licence by IOP Publishing Ltd

Synchronization of semiconductor laser arrays with 2D Bragg structures

V R Baryshev and N S Ginzburg

IAP RAS, 603950, Nizhny Novgorod, Russia

E-mail: baryshev@appl.sci-nnov.ru

Abstract. A model of a planar semiconductor multi-channel laser is developed. In this model two-dimensional (2D) Bragg mirror structures are used for synchronizing radiation of multiple laser channels. Coupling of longitudinal and transverse waves can be mentioned as the distinguishing feature of these structures. Synchronization of 20 laser channels is demonstrated with a semi-classical approach based on Maxwell-Bloch equations.

1. Introduction

Multi-channel laser systems are recognized as a common technique of increasing beam width and power of semiconductor lasers. In those systems the output laser beam is a combination of beams from individual channels. Keeping that combination coherent requires a way of synchronizing the channels. Commonly known methods of synchronization include using external cavities [1], Talbot effect [2] as well as various ways of coupling the channels [3]. In this paper we suggest synchronization of multiple semiconductor heterostructure laser channels with 2D Bragg reflectors. This approach can be effective for planar laser diode arrays and allows integrating the reflectors with the channels on the same substrate as well as using them as an external resonator.

Figure 1 shows the suggested scheme of a multi-channel laser. Active channels with the number of n length of l_z and the width of l_k are separated with air or dielectric lanes with the width of l_d. In the direction of Y axis the channel thickness b_0 is considered sufficient small to allow propagation only one planar waveguide eigenmode, which is typical for distributed feedback (DBF) lasers. Coupling between the channels is provided by special reflectors being planar dielectric waveguides with certain areas covered by 2D Bragg corrugation. The distinguishing feature of those corrugated areas is coupling between longitudinal (Z axis) and transverse (X axis) partial waves [4, 5]. As shown below, this coupling provides mutual synchronization of the laser channels, including channels with slightly different wavelengths.

2. Nonlinear model of a 2D DBF multi-channel laser

A 2D Bragg reflector based on planar dielectric waveguide has rectangular shaped area with the following double periodic sinusoidal modulation of waveguide thickness:

$$b(x,z) = b_0 + b_{1,2}(\cos(\overline{h}(x+z)) + \cos(\overline{h}(x-z))) \tag{1}$$

where $b_{1,2}$ is the modulation amplitudes of the left and right reflectors respectively, $\overline{h} = 2\pi/d$, d is

Content from this work may be used under the terms of the Creative Commons Attribution 3.0 licence. Any further distribution of this work must maintain attribution to the author(s) and the title of the work, journal citation and DOI.
Published under licence by IOP Publishing Ltd

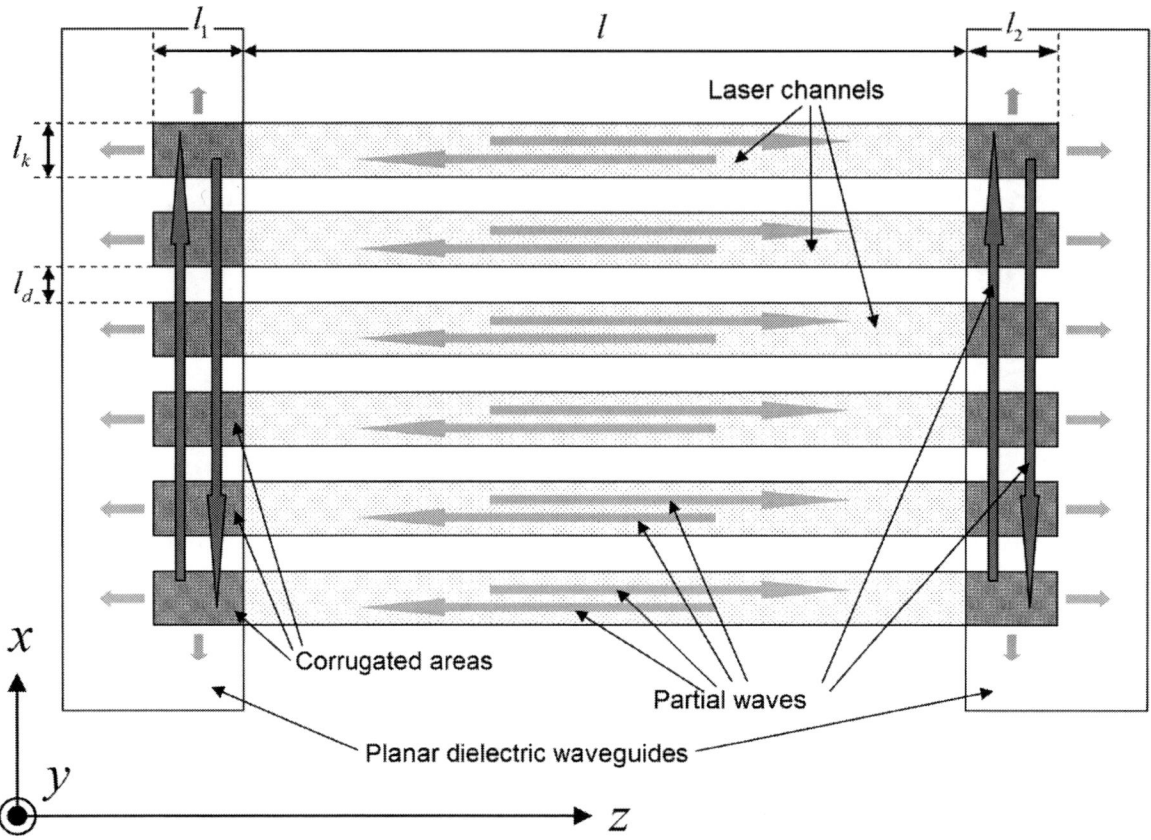

Figure 1. Multi-channel semiconductor laser with 2D Bragg reflectors.

the modulation period along the x- и z- coordinates. Under the Bragg resonance conditions:

$$h \approx \overline{h} \tag{2}$$

those structures provide mutual coupling of the following four partial wave-beams [4, 5]:

$$\vec{A} = \mathrm{Re}\left[\left(\vec{a}_1(y)\left(C_z^+ \mathrm{e}^{-ihz} + C_z^- \mathrm{e}^{ihz}\right) + \vec{a}_2(y)\left(C_x^+ \mathrm{e}^{-ihx} + C_x^- \mathrm{e}^{ihx}\right)\right)\mathrm{e}^{i\omega t}\right] \tag{3}$$

where $\vec{a}_{1,2}(y)$ are eigenwaves of a planar dielectric waveguide propagating along x- и z- directions, $C_{x,z}^\pm(x,z,t)$ are complex amplitudes of partial waves. Let us mention that $C_z^\pm$ waves are amplified when propagate through the laser channels while $C_x^\pm$ waves propagate only in the reflectors and are produced by Bragg scattering. We assume that the channels are wide enough in the scale of Fresnel paremeter ($l_k^2 / l_z \lambda \gg 1$) to neglect the diffraction and assume $C_z^\pm \equiv 0$ at $x \notin (j(l_k + g), j(l_k + g) + l_k)$, where $j = 0...n-1$, which means that waves $C_z^\pm$ exist only in laser channels and the corresponding parts of the reflectors. Similarly, we will consider the transverse waves only inside the reflectors, i.e. $C_x^\pm \equiv 0$ at $z \notin (0, l_1)$ and $z \notin (l + l_1, l + l_1 + l_2)$. It is important to mention that the sinusoidal modulation (1) can be replaced by a "chessboard" modulation and can also be placed inside the structure on the boundary surface of its waveguide layer.

Mutual Bragg scattering of the longitudinal and transverse wave- beams can be described by the following equations:

$$\left(\pm\frac{\partial}{\partial Z}+\frac{\partial}{\partial\tau}\right)C_z^{\pm}+i\hat{\alpha}_{1,2}\left(C_x^{+}+C_x^{-}\right)=0,$$

$$\left(\pm\frac{\partial}{\partial X}+\frac{\partial}{\partial\tau}\right)C_x^{\pm}+i\hat{\alpha}_{1,2}\left(C_z^{+}+C_z^{-}\right)=0,$$

(4)

where $X=x/l_1$, $Z=z/l_1$ and $\tau=tv_g/l_1$ are normalized coordinates. Coupling parameters $\hat{\alpha}_{1,2}$ is given in [5].

Polatization P and inversion ρ of the active media can be represented by components that interact with partial waves $C_z^{\pm}$:

$$P=\mathrm{Re}\left(i\left(P_z^{+}e^{i\bar{h}z}+P_z^{-}e^{-i\bar{h}z}\right)e^{i\omega_0 t}\right),$$

$$\rho=\rho_0+\mathrm{Re}\left(\rho_{2z}e^{2i\bar{h}z}\right),$$

(5)

where $\rho_{2z}(x,z,t)$ is the inversion lattice produced by the spatial hole burning effect.

We will use a semi-classical approach [6] in which the lasing process can be described by the following set of equations:

$$\left(\pm\frac{\partial}{\partial Z}+\frac{\partial}{\partial\tau}\right)\hat{C}_z^{\pm}=\hat{P}_z^{\pm},$$

$$\frac{\partial\hat{P}_z^{+}}{\partial\tau}+\frac{\hat{P}_z^{+}}{\hat{T}_2}+i\hat{\delta}_j=\beta\left(2\hat{C}_z^{+}\hat{\rho}_0+\hat{C}_z^{-}\hat{\rho}_{2z}\right),$$

$$\frac{\partial\hat{P}_z^{-}}{\partial\tau}+\frac{\hat{P}_z^{-}}{\hat{T}_2}+i\hat{\delta}_j=\beta\left(2\hat{C}_z^{-}\hat{\rho}_0+\hat{C}_z^{+}\hat{\rho}_{2z}^{*}\right),$$

$$\frac{\partial\hat{\rho}_0}{\partial\tau}+\frac{(\hat{\rho}_0-1)}{\hat{T}_1}=-\mathrm{Re}\left(\hat{C}_z^{+}\hat{P}_z^{+*}+\hat{C}_z^{-}\hat{P}_z^{-*}\right),$$

$$\frac{\partial\hat{\rho}_{2z}}{\partial\tau}+\frac{\hat{\rho}_{2z}}{\hat{T}_1}=-\left(\hat{C}_z^{+}\hat{P}_z^{-*}+\hat{C}_z^{-}\hat{P}_z^{+*}\right),$$

(6)

Here we use the following normalized variables:

$$\hat{\rho}=\frac{\rho}{\rho_e},\quad \hat{P}_z^{\pm}=P_z^{\pm}\left(\frac{\pi b_\rho l_1}{\rho_e\hbar\omega_0 cv_{gr}b_{eff}}\right)^{\frac{1}{2}},\quad \hat{C}_{x,z}^{\pm}=C_{x,z}^{\pm}\left(\frac{b_{eff}\omega_0}{\pi\rho_e\hbar cv_{gr}b_\rho}\right)^{\frac{1}{2}},$$

$$\hat{T}_{1,2}=\frac{v_{gr}T_{1,2}}{l_1},\quad \beta=\pi\rho_e|\mu|^2 b_\rho l_1 c/2\hbar\omega_0 b_{eff}$$

where ρ_e is the equilibrium concentration of inverted active elements (nonequilibrium carriers in semiconductor laser media) without radiation, μ is the dipole element $T_{1,2}$ are the carrier and polarization relaxation times, δ_j is the detuning of the channel number j middle frequency from the Bragg frequency, b_{eff} is the effective waveguide thickness for the TM waveguide waves (given in [7]), v_{gr} is the group velocity of the amplified waves inside the laser channels, b_ρ is the active layer thickness. All of the structure dimensions are also normalized: $L_2=l_2/l_1$, $L=l/l_1$, $L_k=l_k/l_1$, $L_d=l_d/l_1$.

It is important to emphasize that the equations (6) don't use the balance assumption, which neglects the polarization relaxation time. The transverse relaxation time T_2 acts as the reverse linewidth of the laser media. This complication allows us to take possible frequency mismatches between laser channels into account. In our simulations the channel frequencies δ_j were chosen randomly from an interval with the width of δ in the vicinity of the Bragg frequency: $\delta_j \in (-\delta/2, \delta/2)$.

3. Simulation of a multi-channel laser excitation process
Excitation, synchronization and the steady state regime of a multichannel laser with 2D Bragg reflectors is described by equations (4) and (6) and can be studied numerically Simulation results are

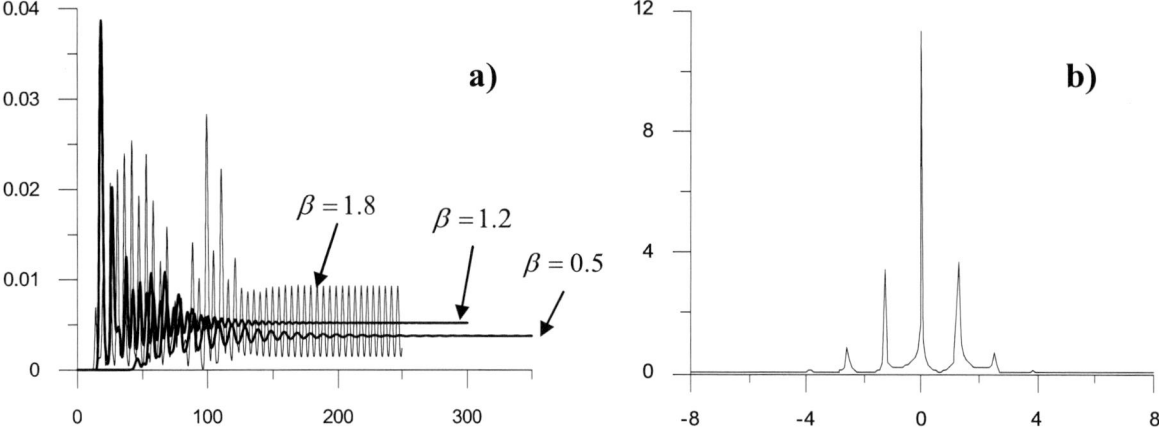

Figure 2. Time dependencies of radiation power at different normalized gain values (a); radiation spectrum in multi-mode regime at $\beta = 1.8$ (b); $n = 20$, $L_2 = 1$, $L = 4$, $L_k = 0.1$, $L_k = 0.05$, $\hat{T}_{1,2} = 1$, $\delta = 1$.

presented in figures 2-4 for the case of 20 laser channels. Establishment of the steady state regime is illustrated in figure 2 by time dependencies of the laser output power. There is a large interval of normalized gain values (parameter β in (6)) where the laser operates in the steady state regime. However, increasing the gain further results in a multimode regime in which the radiation spectrum consist of several longitudinal modes (figure 2b) similar to the Fabry-Perot resonator modes. One can mention that unlike lasers with single section 2D Bragg resonators [4,5], in the considered system single mode excitation can't be provided on the linear stage of the excitation process. The two main reasons for that are presence of multiple longitudinal modes with close quality values and randomness of individual channel frequencies. Accordingly, synchronization of radiation and selection of one longitudinal mode is a result of nonlinear mode interaction.

It is important to emphasize that all four of the partial waves don't have any reflective boundary conditions at the edges of the resonator so the laser radiation goes in all four directions through planar dielectric waveguides (see figure 1). However distribution of the output power between those directions can be made significantly unequal by choosing different length and modulation amplitudes for the left and right reflectors. Stationary distributions of partial waves $C_{x,z}^{\pm}$ are presented in figure 3 for the case where about 80% of radiation power is emitted though $Z = 0$ plane by the C_z^- partial wave.

Space-time distribution of the main C_z^- partial wave phase is presented in figure 4. One can notice that phase distribution in the steady state regime is different inside different channels and seems to be

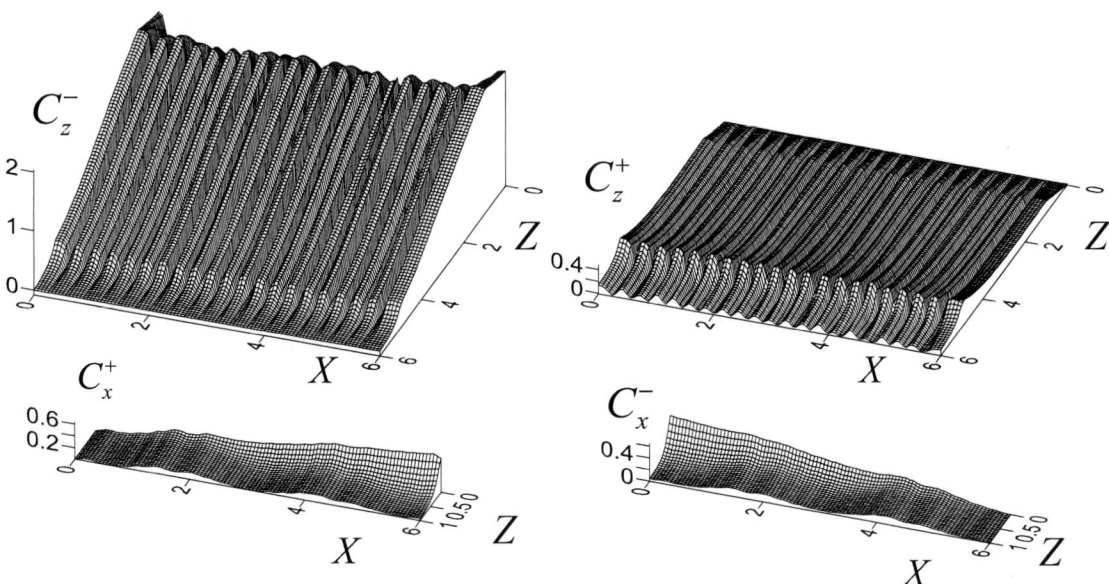

Figure 3. Spatial distributions of partial waves in the steady state regime; $n = 20$, $L_2 = 1$, $L = 4$, $L_k = 0.1$, $L_k = 0.05$, $\hat{T}_{1,2} = 1$, $\beta = 0.5$, $\delta = 1$.

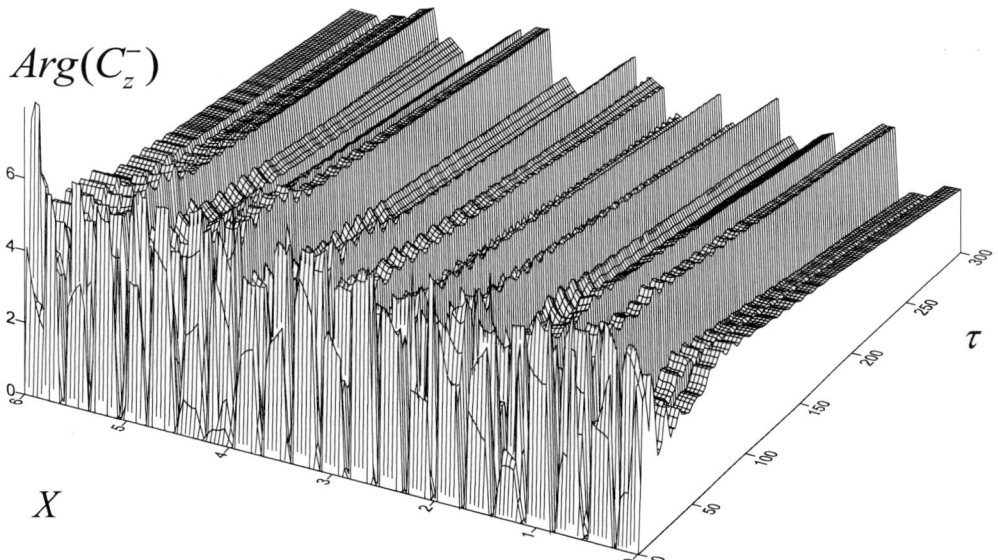

Figure 4. Space-time distribution of the phase of partial wave C_z^- that defines the output radiation structure; $n = 20$, $L_2 = 1$, $L = 4$, $L_k = 0.1$, $L_k = 0.05$, $\hat{T}_{1,2} = 1$, $\beta = 0.5$, $\delta = 1$.

random, which is caused by randomness of the channel frequencies δ_j. In this particular simulation the channel frequencies distribution size δ was comparable to the channel linewidth.

4. Conclusion

2D Bragg structures allow synchronization of multiple semiconductor laser channels when used as external reflectors. Eigenmode spectrum of a resonator with two 2D Bragg structures is located inside the reflection band of the structures. Similarly to a Fabry Perot resonator, the higher quality part of this spectrum consists of equidistant modes with the same transverse but different longitudinal indices. Simulation demonstrates excitation of multiple modes at the initial linear stage of lasing process. At the nonlinear stage the steady state regime establishment corresponding to mutual synchronization of laser array in a wide range of individual channel frequencies and gain values.

Acknowledgements

This work was supported by Competitiveness Program of National Research Nuclear University MEPhI.

References

[1] Lui B *et al* 2014 *Optics Communications* **324** 301
[2] Goldobin I S 1989 *Soviet Journal of Quantum Electronics* **19**(10) 1261
[3] Peleš S, Rogers J L and Wiesenfeld K 2006 *Phys. Rev. E* **73** 026212
[4] Baryshev V R, Ginzburg N S, Malkin A M and Sergeev A S 2009 *Quantim Electron* **39**(12) 1159
[5] Ginzburg N S, Baryshev V R, Sergeev A S and Malkin A M 2015 *Phys. Rev. A* **91** 053806
[6] Andreev A V 1990 *Sov. Phys. Usp.* **33** № 12 997
[7] Kogelnik H and Shank C V 1971 *Appl. Phys. Lett.* **18** № 4 152

Theory of operating characteristics of a semiconductor quantum well laser: Inclusion of global electroneutrality in the structure

Z N Sokolova[1], N A Pikhtin[1], I S Tarasov[1] and L V Asryan[2]

[1] Ioffe Institute, St. Petersburg, 194021, Russia
[2] Virginia Polytechnic Institute and State University, Blacksburg, VA 24061, USA

E-mail: zina.sokolova@mail.ioffe.ru, asryan@vt.edu

Abstract
A model for calculating the operating characteristics of semiconductor quantum well (QW) lasers is presented. The model exploits the condition of global electroneutrality, which includes the charge carriers both in the two-dimensional (2D) active region (QW) and bulk waveguide region (optical confinement layer – OCL). The charge of each sign in the OCL is shown to be significantly larger than that in the QW. As a result of this, (i) the global electroneutrality condition reduces to the condition of electroneutrality in the OCL and (ii) the local electroneutrality in the QW can be strongly violated, i.e., the 2D electron and hole densities in the QW can significantly differ from each other.

1. Introduction
The stimulated emission in contemporary semiconductor lasers is generated in a low-dimensional active region, which is surrounded by a wider band gap bulk waveguide region (optical confinement layer – OCL) [1, 2]. In such structures, electrons and holes are first injected from the cladding layers into the OCL and then captured into the active region [3] (figure 1).

In [4], the operating characteristics of semiconductor QW lasers were calculated with a proper account for noninstantaneous capture of charge carriers. In calculations, we assumed the local electroneutrality in QWs, i.e., equality of the 2D electron and hole densities in QWs to each other.

In this work, the operating characteristics of a semiconductor QW laser are calculated exploiting the condition of global electroneutrality, which includes the charge carriers both in the active region (QWs) and OCL. This condition thus presents the equality of the total charge of electrons to that of holes – see equation (10) below. The charge of each sign in the OCL is shown to be significantly larger than that in the QW. As a result of this, (i) the global electroneutrality condition reduces to the condition of electroneutrality in the OCL and (ii) the local electroneutrality in the QW can be strongly violated, i.e., the 2D electron and hole densities in the QW can significantly differ from each other.

Content from this work may be used under the terms of the Creative Commons Attribution 3.0 licence. Any further distribution of this work must maintain attribution to the author(s) and the title of the work, journal citation and DOI.

Published under licence by IOP Publishing Ltd

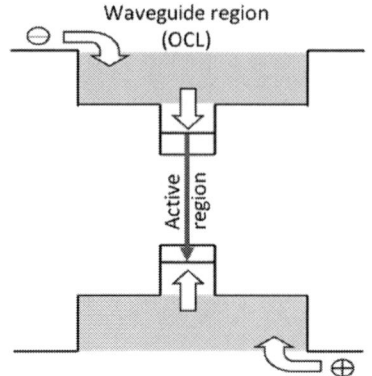

Figure 1. A schematic of a semiconductor laser with a quantum-confined active region.

2. Theoretical model

To calculate the laser characteristics, the following set of steady-state rate equations is used [4]:

for free electrons in the bulk OCL [$b(\partial n^{OCL}/\partial t) = 0$],

$$\frac{j}{e} + N_{QW}\frac{n^{QW}}{\tau_{n,esc}} - N_{QW}v_{n,capt,0}(1-f_n)\,n^{OCL} - bB_{3D}n^{OCL}p^{OCL} = 0 \tag{1}$$

for free holes in the OCL [$b(\partial p^{OCL}/\partial t) = 0$],

$$\frac{j}{e} + N_{QW}\frac{p^{QW}}{\tau_{p,esc}} - N_{QW}v_{p,capt,0}(1-f_p)p^{OCL} - bB_{3D}n^{OCL}p^{OCL} = 0 \tag{2}$$

for electrons confined in the QW ($\partial n^{QW}/\partial t = 0$),

$$v_{n,capt,0}\,(1-f_n)\,n^{OCL} - \frac{n^{QW}}{\tau_{n,esc}} - B_{2D}n^{QW}p^{QW} - v_g\,g^{max}(f_n + f_p - 1)\frac{N}{S} = 0 \tag{3}$$

for holes confined in the QW ($\partial p^{QW}/\partial t = 0$),

$$v_{p,capt,0}(1-f_p)p^{OCL} - \frac{p^{QW}}{\tau_{p,esc}} - B_{2D}n^{QW}p^{QW} - v_g\,g^{max}(f_n + f_p - 1)\frac{N}{S} = 0 \tag{4}$$

and for photons in the lasing mode ($\partial N/\partial t = 0$),

$$v_g N_{QW}\,g^{max}(f_n + f_p - 1)N - v_g(\beta + \alpha_{int})N = 0 \tag{5}$$

The following quantities are the unknowns in equations (1)-(5): n^{OCL} and p^{OCL} are the free electron and hole densities in the OCL, n^{QW} and p^{QW} are the 2D densities of electrons and holes confined in each of the QWs, N is the number of photons in the stimulated emission, f_n (f_p) is the occupancy of the lower (upper) edge of the electron (hole) quantum-confinement subband in the QW. The occupancies f_n and f_p are expressed in terms of the 2D electron and hole densities n^{QW} and p^{QW} as follows [7, 8]:

$$f_n = 1 - \exp\left(-\frac{n^{QW}}{N_c^{2D}}\right), \qquad f_p = 1 - \exp\left(-\frac{p^{QW}}{N_v^{2D}}\right) \tag{6}$$

where $N_{c,v}^{2D} = m_{e,hh}^{QW} T / (\pi \hbar^2)$ are the 2D effective densities of states in the conduction and valence bands in the QW, $m_{e,hh}^{QW}$ are the electron and hole effective masses in the QW, and the temperature T is measured in units of energy.

The following parameters enter into equations (1)-(5): j is the injection current density, e is the electron charge, N_{QW} is the number of identical (of the same material composition and width) QWs, $\tau_{n,esc}$ and $\tau_{p,esc}$ are the thermal escape times of electrons and holes from a QW to the OCL, $v_{n,capt,0}$ and $v_{p,capt,0}$ are the capture velocities of electrons and holes into an empty (at $f_n = 0$ and $f_p = 0$) QW measured in units of cm/s, b is the thickness of the OCL, B_{3D} and B_{2D} are the spontaneous radiative recombination coefficients for the bulk (OCL) and 2D (QW) regions measured in units of cm^3/s and cm^2/s, respectively – see [5] and [6] for the expressions for B_{3D} and B_{2D}; v_g is the group velocity of light, g^{max} is the maximum gain in each QW, $S = WL$ is the cross-section of the junction, W is the lateral size of the device, L is the Fabry-Pérot cavity length, $\beta = (1/L)\ln(1/R)$ is the mirror loss, R is the facet reflectivity, and α_{int} is the coefficient of internal optical loss in the structure.

The thermal escape times of electrons and holes from the QW into the OCL are [4, 9]:

$$\tau_{n,\,esc} = \frac{1}{v_{n,\,capt,0}(1-f_n)} \frac{N_c^{2D}}{n_1}, \qquad \tau_{p,\,esc} = \frac{1}{v_{p,\,capt,0}(1-f_p)} \frac{N_v^{2D}}{p_1} \tag{7}$$

Where

$$n_1 = N_c^{3D} \exp\left(-\frac{\Delta E_c - \varepsilon_n^{QW}}{T}\right), \qquad p_1 = N_v^{3D} \exp\left(-\frac{\Delta E_v - \varepsilon_p^{QW}}{T}\right) \tag{8}$$

and $N_{c,v}^{3D} = 2\left[m_{c,v}^{OCL} T / (2\pi \hbar^2)\right]^{3/2}$ are the 3D effective densities of states in the conduction and valence bands in the OCL, $m_{c,v}^{OCL}$ are the electron and hole effective masses in the OCL, $\Delta E_{c,v}$ are the conduction and valence band offsets at the heterointerface of the QW and OCL, and ε_n^{QW} (ε_p^{QW}) is the energy of the lower (upper) edge of the electron (hole) subband in the QW.

The velocities $v_{n,capt,0}$ and $v_{p,capt,0}$ of electron and hole capture from the OCL into an empty QW are the characteristics of the QW; they thus depend on the QW width and depth, i.e., on the material compositions of the QW and surrounding layers. The capture velocity can be significantly different in different laser structures. In [10], we determined the value of the electron capture velocity in the laser structure similar to the one used here as an example for our calculations (see below).

The total capture velocities $v_{n,capt}$ and $v_{p,capt}$ are defined as [4]

$$v_{n,\,capt} = v_{n,\,capt,0}(1-f_n), \qquad v_{p,\,capt} = v_{p,\,capt,0}(1-f_p). \tag{9}$$

The condition of global electroneutrality for the OCL and QW is written as follows:

$$e\left(N_{QW}\, n^{QW} + b n^{OCL}\right) = e\left(N_{QW}\, p^{QW} + b p^{OCL}\right) \tag{10}$$

The set of equations (1)-(5), added by equation (10), was solved numerically for a InGaAs/GaAs/AlGaAs laser structure containing a single strained In$_{0.28}$Ga$_{0.72}$As QW of 80 Å width. The material of the OCL of thickness $b = 1.7$ μm was GaAs, the material of the cladding layers was $Al_{0.3}Ga_{0.7}As$. The lasing wavelength was 1.044 μm. The Fabry-Pérot cavity length $L = 1.5$ mm, the stripe width $W = 100$ μm, the mirror reflectivity $R = 0.32$, the mirror loss $\beta = 7.6$ cm^{-1}, the temperature $T = 300$ K, the coefficient of internal optical loss $\alpha_{int} = 1$ cm^{-1} (in this work, α_{int} is assumed to be independent of the injection current). The maximum modal gain in the QW $g^{max} = 49.1$ cm^{-1}. The following values were used for the electron and hole capture velocities [10]: $v_{n,capt,0} = 4 \times 10^5$ cm/s and $v_{p,capt,0} = 4 \times 10^5$ cm/s.

3. Discussion of laser characteristics

The following characteristics were calculated versus the injection current density j (up to $j = 75$ kA/cm^2): i) electron and hole densities in the QW, n^{QW} and p^{QW} (figure 2), ii) electron and hole densities in the OCL, n^{OCL} and p^{OCL} (figure 4), iii) stimulated recombination current density in the QW, j_{stim}, and spontaneous recombination current density in the OCL, j_{spon}^{OCL} (figure 8), and iv) output optical power (figure 9).

The above listed characteristics, calculated using the global electroneutrality condition (10), were compared with those calculated using the condition of local electroneutrality in the QW

$$e\, n^{QW} = e\, p^{QW}. \tag{11}$$

As seen from figure 2, the electron and hole densities in the QW, calculated using the global electroneutrality condition (10), differ significantly.

However, the product $n^{QW}p^{QW}$, calculated with the use of the condition (10), remains virtually constant with increasing injection current (figure 3).

The electron and hole densities in the OCL are shown in figure 4.

As seen from figure 5, the charge of each sign in the OCL is significantly larger than that in the QW,

$$ebn^{OCL} \gg eN_{QW}n^{QW}, \qquad ebp^{OCL} \gg eN_{QW}p^{QW} \tag{12}$$

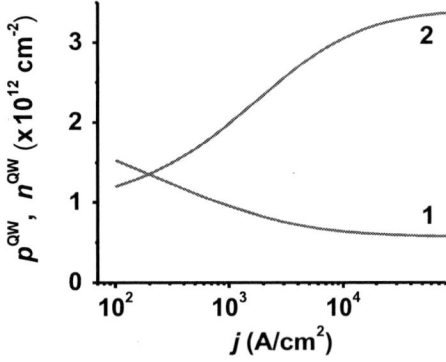

Figure 2. Electron (1) and hole (2) densities in the QW against injection current density calculated using the global electroneutrality condition.

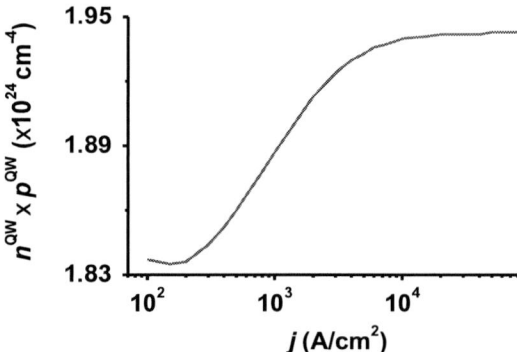

Figure 3. Product of electron and hole densities in the QW against injection current density calculated using the global electroneutrality condition.

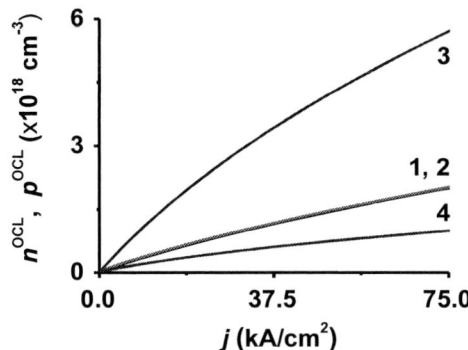

Figure 4. Electron (1, 3) and hole (2, 4) densities in the OCL against injection current density: (1, 2) - using the global electroneutrality condition; (3, 4) - using the local electroneutrality condition in the QW.

Figure 5. Ratio of the electron charge in the OCL to that in the QW (curve 1) and ratio of the hole charge in the OCL to that in the QW (curve 2) against injection current density calculated using the global electroneutrality condition.

In view of inequalities (12), the condition (10) of global electroneutrality reduces to the condition of electroneutrality in the OCL, i.e., i.e., to the equality of the free electron and hole densities in the OCL (figure 4, curves 1 and 2),

$$n^{\text{OCL}} = p^{\text{OCL}} \tag{13}$$

In contrast, the electron and hole densities in the OCL, calculated using the condition (11) of local electroneutrality in the QW, differ considerably (figure 4, curves 3 and 4).

Using in (13) expressions (A27) and (A28) of [4] for n^{OCL} and p^{OCL} and considering high injection current densities, the following relationship can be derived:

$$v_{\text{n,capt,0}}(1 - f_{\text{n}}) = v_{\text{p,capt,0}}(1 - f_{\text{p}}) \tag{14}$$

which is the equality of the total capture velocity of electrons to that of holes at high injection currents.

It follows from (14) that, at equal electron and hole capture velocities into an empty QW ($v_{\text{n,capt,0}} = v_{\text{p,capt,0}}$) and no matter what is their common value, $f_{\text{n}} = f_{\text{p}}$; in the laser structure considered here, $f_{\text{n}} = f_{\text{p}} = 0.588$ (figure 6). If, however, $v_{\text{n,capt,0}} \neq v_{\text{p,capt,0}}$, then, as seen from (14), the level occupancies f_{n} and f_{p} tend to different constant values with increasing injection current (figure 7).

The spontaneous radiative recombination current density in the OCL calculated using the expression

$$j_{\text{spon}}^{\text{OCL}} = ebB_{3\text{D}}n^{\text{OCL}}p^{\text{OCL}} \tag{15}$$

is shown in figure 8 as a function of the injection current density (curves 1 and 2). The spontaneous recombination in the OCL is a parasitic processes, which reduces the quantum efficiency and output optical power of the laser. As seen from the figure, $j_{\text{spon}}^{\text{OCL}}$, calculated using the global electroneutrality condition (curve 1), is lower than $j_{\text{spon}}^{\text{OCL}}$, calculated using the local electroneutrality condition in the QW (curve 2).

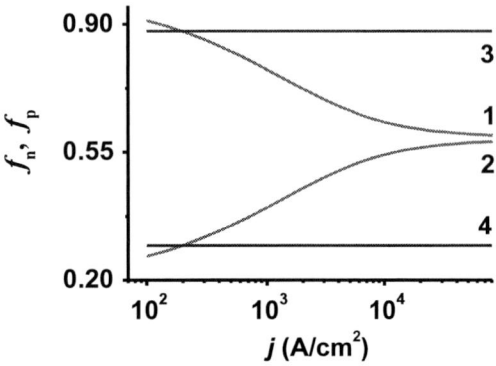

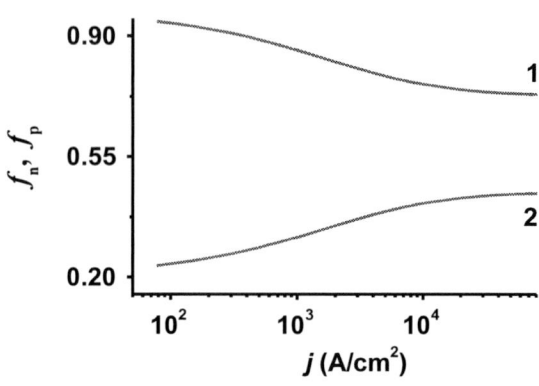

Figure 6. Electron (1, 3) and hole (2, 4) level occupancies in the QW against injection current density: (1, 2) - using the global electroneutrality condition; (3, 4) - using the local electroneutrality condition in the QW. The electron and hole capture velocities into an empty QW $v_{n,capt,0} = v_{p,capt,0} = 4 \times 10^5$ cm/s.

Figure 7. Electron (1) and hole (2) level occupancies in the QW against injection current density calculated using the global electroneutrality condition. The electron and hole capture velocities into an empty QW $v_{n,capt,0} = 10^6$ cm/s and $v_{p,capt,0} = 5 \times 10^5$ cm/s.

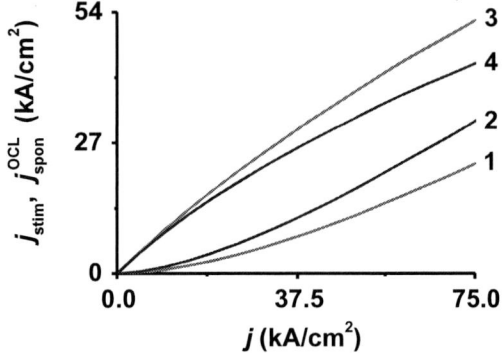

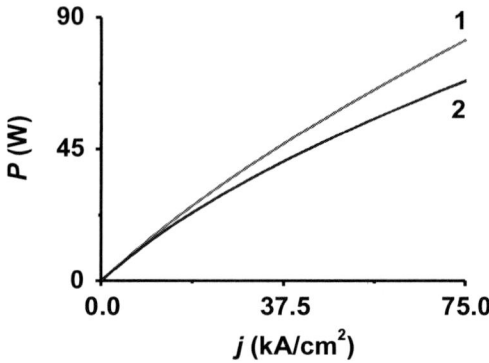

Figure 8. Spontaneous recombination current density in the OCL (1, 2) and stimulated recombination current density in the QW (3, 4) against injection current density: (1, 3) - using the global electroneutrality condition, (2, 4) - using the local electroneutrality condition in the QW.

Figure 9. Optical power of the laser against injection current density (light-current characteristic): (1) - using the global electroneutrality condition, (2) - using the local electroneutrality condition in the QW.

The current density of stimulated radiative recombination in the QW,

$$j_{stim} = e\, v_g (\beta + \alpha_{int}) \frac{N}{S} \tag{16}$$

is also shown in figure 8. The current density of stimulated recombination, calculated using the global electroneutrality condition (curve 3), is higher than that, calculated using the local electroneutrality condition in the QW (curve 4).

The output optical power P against injection current density,

$$P(j) = \frac{\hbar\omega}{e} S\, j_{\text{stim}}(j) \frac{\beta}{\beta + \alpha_{\text{int}}} \qquad (17)$$

is shown in figure 9. The calculation using the global electroneutrality condition (curve 1) gives higher values of the power as compared to that, which uses the local electroneutrality condition in the QW (curve 2).

4. Conclusions

A model for calculating the operating characteristics of semiconductor QW lasers has been presented. The model exploits the condition of global electroneutrality, which includes the charge carriers both in the 2D active region (QW) and bulk waveguide region (OCL). The charge of each sign in the OCL has been shown to be significantly larger than that in the QW. As a result of this, (i) the global electroneutrality condition reduces to the condition of electroneutrality in the OCL and (ii) the local electroneutrality in the QW can be strongly violated, i.e., the 2D electron and hole densities in the QW can significantly differ from each other. The product of the 2D electron and hole densities in the QW, however, remains virtually constant with increasing injection current.

The use of the global electroneutrality condition gives higher values of the power than the use of the local electroneutrality condition in the QW.

Acknowledgments

This work was supported by the federal program of the Ioffe Institute. Z. N. Sokolova is grateful for the hospitality of MEPhI during her visit supported by the Competitiveness Program of National Research Nuclear University MEPhI. L. V. Asryan acknowledges the U.S. Army Research Office (Grant No. W911NF-13-1-0445) for support of this work.

References

[1] Zory P S 1993 *Quantum Well Lasers* ed P S Zory Jr (Academic, Boston) p 504
[2] Kapon E, Ed. 1999 *Semiconductor Lasers I: Fundamentals* (1st Edition, Academic, New York) p 453
[3] Asryan L V, Luryi S and Suris R A 2002 *Appl. Phys. Lett.* **81** 2154
[4] Asryan L V and Sokolova Z N 2014 *J. Appl. Phys.* **115** 023107
[5] Asryan L V and Suris R A 1996 *Semicond. Sci. Technol.* **11** 554
[6] Asryan L V 2005 *Quantum Electron.* **35** (12) 1117
[7] Vahala K J and Zah C E 1988 *Appl. Phys. Lett.* **52** 1945
[8] Asryan L V and Luryi S 2003 *Appl. Phys. Lett.* **83** 5368
[9] Han D S and Asryan L V 2010 *Nanotechnology* **21** 015201
[10] Sokolova Z N, Bakhvalov K V, Lyutetskiy A V, Pikhtin N A, Tarasov I S and Asryan L V 2015 *Electron. Lett.* **51** 780

Diode lasers with front surface high-order distributed Bragg reflector

V V Zolotarev, A Yu Leshko, Z N Sokolova, Ya V Lubyanskiy, N A Pikhtin, D N Nikolaev, V V Shamakhov and I S Tarasov

Ioffe Institute, 194021, St. Petersburg, Russia

E-mail: zolotarev.bazil@mail.ioffe.ru

Abstract. Control of spectral and spatial parameters of semiconductor lasers is a key element of their development and application. Surface integrated high-order distributed Bragg reflector (DBR) is the way to narrow diode laser spectrum preserving the conventional postgrowth process of laser diodes. On the other hand, high-order DBR has specific optical losses, which reduce optical output power. In this work we analyzed these specific optical losses and present design of laser resonator containing short front DBR, that allow to significantly reduce losses. With this approach, slope efficiency was greatly increased in contrast with conventional rear DBR laser diodes and 4 W optical output power with spectral width of less than 3 Å was attained.

1. Introduction

Nowadays semiconductors lasers with distributed Bragg reflector (DBR) is a key element of photonics integrated circuits. The ability of engineering of spectral and spatial behavior of lasers emission allow to significantly widen the application field. Current approach seeks to simplify postgrowth processing of diode lasers with distributed Bragg elements. In this concept the idea of integrated surface high-order DBR looks the most advantageous [1,2]. Such design of resonator may be formed by relatively simple technological processes of photolithography and reactive ion etching. DBR location in the upper cladding layer of heterostructure eliminates the process of epitaxial overgrowth.

This paper is devoted to study of the semiconductor lasers with high-order DBR characteristics. Experimental samples are based on separate-confinement double heterostructure InGaAs/AlGaAa/GaAs, emitting in 1000-11000 nm wavelength range.

Specific features of high-order DBR lasers, which is new sort of optical losses, were investigated. The mechanism of new type of optical losses, which limits output power, is described. The alternative design of diode lasers resonator utilizing DBR as a front mirror is presented. This design decreases specific optical losses and narrows the spectrum. With this approach, we increase slope efficiency of power-light characteristic of laser by three times for nonoptimal DBR devices. Output power was increased up to 4 W. Spectral widths was at least 3 Å over the whole current range.

2. Specific optical losses in high-order surface DBR

Utilizing of simple and conventional technological process of photolithography and reactive ion etching requires to increase DBR period. According to the Bragg law period increase leads to the growth of diffraction order:

$$\Lambda = \frac{N\lambda}{2n} \tag{1}$$

where Λ - DBR period, λ/n - wavelength in medium, N - diffraction order (natural number). Increasing of order result in emergence of a new direction of constructive interferes of emission, diffracted at edge of each groove [3]. Electromagnetic wave running in planar dielectric waveguide with periodical variation of dielectric index generates modes that propagate at the angle to the plain of waveguide according to:

$$\Phi_i = 90° + \arcsin(\frac{2i}{N} - 1), \ i=0,1,2,3...N \tag{2}$$

For N>1 high-order interfering radiation modes (IRMs) are appeared. These IRMs are the parasitic optical output losses, emitting from the surface of DBR. If we neglect the internal optical losses in heterostructure, the losses corresponding to the interfering high-order modes could be calculated as 1-R-T, where R and T are reflection and transmission of the DBR. In this work we used couple mode theory describing the behavior of light in waveguide with weak periodic variation of optical parameter. This theory was presented by Kogelnik and Shank [4] and Yariv [5].

According to the theory under consideration, two resonance modes exchange energy determined by coupling coefficient:

$$\kappa = \frac{k_0^2}{2\beta} \int \Delta\varepsilon_N(x) \cdot U^2(x) dx \tag{3}$$

k0 – wavenumber in air, β - wavenumber in media (complex value), U(x) - normalized profile of transvers guided mode (TEM$_0$), $\Delta\varepsilon$N- N-Fourier coefficient of periodic function of permittivity expansion. Coupling coefficient determines energy exchange value. Fourier coefficient presence determines coupling coefficient (and reflection coefficient as well) dependence on geometric shape of the groove of the Bragg grating.

The dependence of the reflection on geometric shape of the groove is demonstrated on figure 1. The shape of the groove is schematically shown in the inset figure. According to the results of calculation, we can observe almost complete reduction of reflection during the transition from V-shaped groove to the rectangular-shaped. However, transmission increasing is not observed, indicating on parasitic optical losses increase corresponding to the IRMs. Thus, we can conclude, taking into account the potential of reactive ion etching technology, that the V-shape is the most optimal form of Bragg grating groove. However theoretical calculation shows, that even this optimal form does not allow to completely avoid IRMs optical losses and 1-R-T≠0.

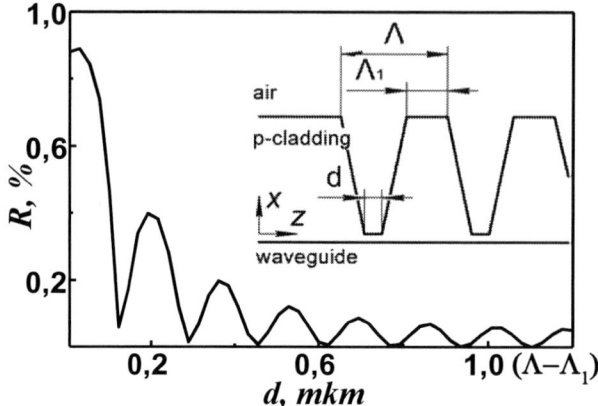

Figure 1. The dependence of the reflection on geometric shape of the Bragg grating groove.

Dependences of R, T, 1-R-T on DBR length are presented on figure 2. The effective length (saturation length) is determined by coupling coefficient. Reflectance (R) and losses IRMs (1-R-T) saturation value are determined by geometric shape of groove. However, it is clear that for any geometrical shape shorter grating provides smaller losses related to IRMs. Also, reducing the length causes reflectance reduction and DBR transmission increase. In the case of fabricating experimental laser diode samples with nonoptimal groove geometrical shape which are characterized by high rate of parasitical optical losses, the only solution to obtain high optical power is laser with front DBR. Reducing the length of the grating will significantly reduce losses of IRMs in the cavity and therefore significantly increase optical power of the laser.

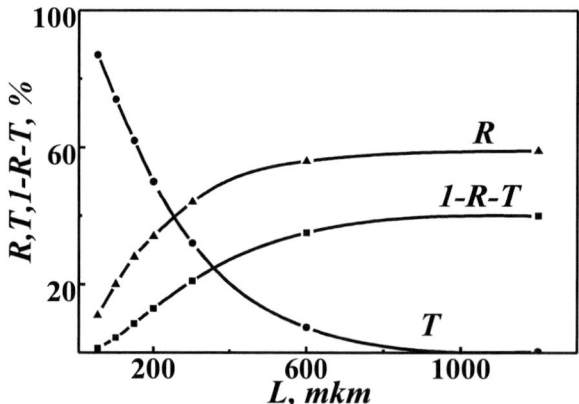

Figure 2. The dependence of reflectance, transmission coefficients and optical losses of IRMs on length of DBR.

In addition to typical high-order IRMs losses there are two more loss mechanism in integrated surface DBR. Since DBR is electrical passive, there is no gain in this section of cavity. At the same time the internal optical losses related to electromagnetic waves scattering on defects and dopant remain. Thus, in the case of light penetration in DBR on depth L_{eff}, reflected light intensity diminishes according to the Buger law:

$$I = I_0 \cdot R \cdot \exp(-2\alpha_i L_{eff}) \qquad (4)$$

I_0 and I – intensity of the incident and reflected light, respectively, α_i – internal optical loss in heterostructure. Modern separate-confinement double heterostructure InGaAs/AlGaAa/GaAs are characterized by very small value [6] of α_i, but it is still necessary to take into account its contribution.

The second mechanism is associated with the etched surface roughness – the technological feature of reactive ion etching. Presence of roughness on the facets of the grating groove also leads to the light scattering. The magnitude of these losses is also determined by the length of way, that light passed through DBR.

All of the three abovementioned optical loss mechanism depend on the length of DBR. Increasing the laser power requires a reduction of optical losses. According to the provided calculations, solution of problem under consideration is construction of laser resonator with short DBR as a front mirror with R≤20%. In this case it is necessary to deposit antireflection coating front facet of laser, and high reflection coating (R=99%) rear facet.

3. Semiconductor laser with a front DBR

Experimental samples of semiconductor lasers with integrated high-order surface DBR based on InGaAs/Al$_{0.1}$Ga$_{0.9}$Aa/Al$_{0.25}$Ga$_{0.75}$As heterostructures (figure 3) have been fabricated. DBR period was Λ=2.4 µm. Diffraction order – N=16. Wavelength – λ=1028 nm. The samples didn't have optimal DBR parameters. Grating groove geometric pattern was d=0.35µm that produced parasitic loss of 50% differential efficiency [7]. Three types of samples have been studied. The first group consisted of lasers with 1 mm length rear DBR which was a totally reflecting mirror. The front facet was low-reflection coated down to R=5%. Two other groups included semiconductor lasers with 0.5 mm and 0.25 mm DBR that were front mirrors. The samples differed only by DBR length. The rear facet was high-reflectance coated up to R>95%. The front facet was antireflection coated (AlN) down to R<0.5% in order to sufficiently increase Fabry-Perot mode threshold conditions.

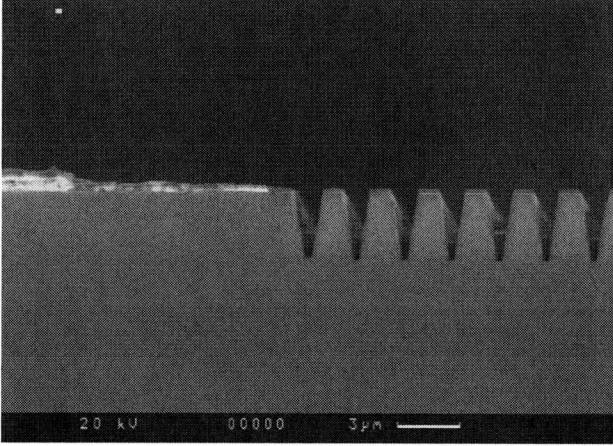

Figure 3. SEM image of a surface high-order DBR.

Light-current characteristics of the samples are shown on the figure 4. DBR length decrease leads to output optical power increase. An increase in external differential quantum efficiency gives evidence of optical loss decrease in the laser cavity. It can be seen that even in spite of nonoptimal DBR implementation one can get high values of optical power. Spectra set of the sample which had demonstrated the highest slope of a light-current characteristic is shown on the figure 5. Spectrum structure is multimode but it's width does not exceed 3 Å in the whole range of pump currents. Wavelength shift is due to temperature induced permittivity change [8].

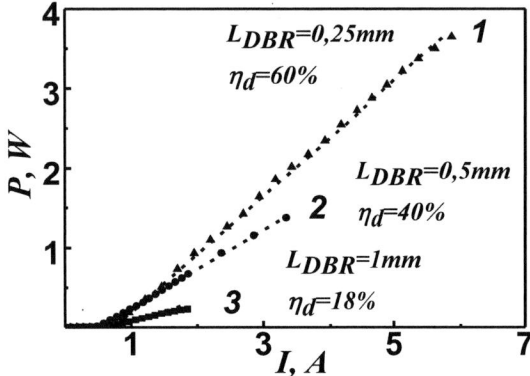

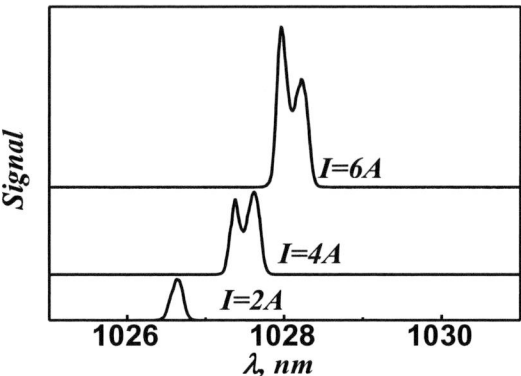

Figure 4. Light-current characteristics of the three sample types. 1, 2 – front DBR of L=250mkm, 500mkm respectively, 3 – DBR rear facet of L=1000mkm.

Figure 5. Spectra of a semiconductor laser with a front DBR of 0.25 mm length.

4. Conclusion

Surface high-order distributed Bragg mirror provides a semiconductor laser spectrum narrowing. However, the DBR also has specific optical loss that limits output optical power. Loss mechanisms deal with high-order interfering radiation modes, lack of gain in the DBR section and accidental light scattering on the groove roughness. Loss value for each case is proportional to DBR length.

Therefore, DBR length decrease leads to optical loss reduction in the cavity, reflection coefficient decrease and light transmittance increase. Use of DBR as a front mirror allows one to narrow a semiconductor laser spectrum keeping at the same time a high value of output optical power. Even in the case of nonoptimal DBR groove geometrical pattern that results in gigantic loss of higher-order IRMs the presented laser cavity design approach allowed reaching of 60 % external differential quantum efficiency and 4 W optical power at spectrum width of less than 3 Å.

Acknowledgments

This work was supported by the federal program of the Ioffe Institute. Z. N. Sokolova is grateful for the hospitality of MEPhI during her visit supported by the Competitiveness Program of National Research Nuclear University MEPhI.

References

[1]   Price R K, Elarde V C and Coleman J J 2007 *J. Appl. Phys.* **101** 053116
[2]   Fricke J, Bugge F, Ginolas A, John W, Klehr A, Matalla M, Ressel P, Wenzel H and Erbert G 2010 *IEEE Photonics Technol. Lett.* **22**(5) 284

[3]   Casey H C and Panish M B 1978 Heterostructure lasers: Fundamental principles (Academic Press)
[4]   Kogelnik H and Shank C V 1972 *J. Appl. Phys.* **43** 2327–35
[5]   Yariv A 1973 *IEEE J. Quantum Electron.* **QE-9** 919–33
[6]   Pikhtin N A, Slipchenko S O, Sokolova Z N, Stankevich A L, Vinokurov D A, Tarasov I S and Alferov Zh I 2004 *Electronics Letters* **40**(22) 1413
[7]   Zolotarev V V, Leshko A Yu, Pikhtin N A, Slipchenko S O, Sokolova Z N, Lubyanskiy Ya V, Voronkova N V and Tarasov I S 2015 *QUANTUM ELECTRON* **45**(12) 1091–7
[8]   Zolotarev V V, Leshko A Yu, Pikhtin N A, Lyutetskiy A V, Slipchenko S O, Bakhvalov K V, Lubyanskiy Ya V, Rastegaeva M G and Tarasov I S 2014 *QUANTUM ELECTRON* **44**(10) 907–11

The peculiarities of spectra in high power 970 laser diodes

V V Bezotosnyi [1,2], O N Krokhin[1,2], V A Oleshchenko [2], V F Pevtsov [2], Yu M Popov,[1,2] and E A Cheshev [1,2]

[1]National Research Nuclear University MEPhI, Kashirskoe shosse 31, 115409, Moscow, Russia
[2] P.N. Lebedev Physical Institute of the Russian Academy of Sciences, Leninsky Prospect 53, 119991, Moscow, Russia

E-mail: victorbe@sci.lebedev.ru, vvbezotosniy@mephi.ru

Abstract. We present the results of optical power, total efficiency and spectra of CW laser diodes emitting at wavelength 970 nm. Total efficiency in maximum 72 % was measured and reliable operation at 15 W CW was achieved. At the base of measured CW spectral parameters in wide range of pump currents we discuss the possible reasons of observed features in dependences of spectral envelope, spectral maximum and spectral half-width against pumping current.

1. Introduction

High-power laser diodes are still in focus of researchers and engineers who develop different types of the optoelectronic devices. The areas of their applications quickly expands in technology, navigation, medicine and other spheres of science and technique. Taking that in account the increase of reliable and maximum operating optical power and also the brightness of laser diodes becomes more and more actual [1–6]. Several research groups published the results describing the achievements of 20-25 W range CW output power. At the same time the best laser diodes available at market have about 10 W CW power in spectral range 800-1060 nm and much efforts must be applied to get commercial products with 20 W CW power. Spectral features of such devices at high power operation are not well examined and still the object of scientific and practical interest.

2. Optical power and total efficiency

The most important problem for high-power laser diodes is efficient cooling of active region of the laser crystal. To increase the output power and to receive the reliable operation we need to ensure the several important parameters under the assembling of the laser chip at heat conducting elements. First of all we need to make the heat transfer extremely uniform at heat transfer interface and also to prevent the appearance of excess thermo-elastic strain in laser crystal which appear under assembling via the difference of thermal expansion coefficients of semiconductor material and thermo-conducting heatsinks. The experimental dependence of the output optical power and total efficiency regarding the pumping current for laser chip with stripe contact width 95 μm and resonator length 4 mm is presented at figure 1.

Content from this work may be used under the terms of the Creative Commons Attribution 3.0 licence. Any further distribution of this work must maintain attribution to the author(s) and the title of the work, journal citation and DOI.
Published under licence by IOP Publishing Ltd

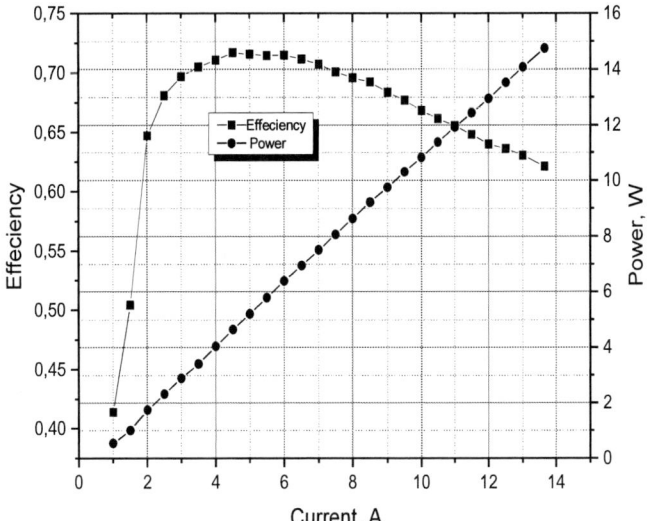

Figure 1. The dependence of output power and total efficiency regarding the pumping current for laser diode emitting at 970 nm assembled on F-mount type heat-sink in CW operation under heat-sink temperature 20 °C.

Threshold current of laser was 0.9 A, maximum total efficiency was observed at pumping current around 5 A. Average slope of W/ A characteristic 1.12 W/A was measured in pumping current range 1.5 – 14 A. Laser operation was stable at output power 15.1 W under 120 hours without output power change inside the range of power measurement accuracy.

3. Spectral parameters

Spectral map of laser radiation is quite important for several practical applications, particularly the spectral position of spectral envelope maximum and the half- width of spectral envelope are important parameters for matching the spectra of pumping diode laser radiation with absorption spectra of solid-state active media.

The spectral parameters of high power laser diode in current range 1-13 A with 1 A step are presented at figure 2.

The important feature of spectra at figure 2 is the presence of some peculiarities at currents more than 5 A. At currents below 5 A spectral envelope is quite smooth with one maximum and slight asymmetry on the short wavelength side, particularly it have the more long tail on the envelope. By the way at 5 A we observe the maximum of total efficiency. At pumping currents more than 5 A the spectral envelope contains not only single maximum yet. At short wavelength side of the envelope we observe the new peculiarities which appears first as two and later as three competition peaks. At pumping currents more than 6 A the maximum of spectral envelope contains of 3 peaks, and with the increase of pumping current the intensity of these peaks at some currents becomes almost equal. To gain the correct knowledge regarding the appearance of such spectral peculiarities we need to conduct additional research of spacial distribution of radiation in near and far field at the lot of uniform samples.

One of the probable causes of such changes in spectral distribution of radiation with the increase of pumping current can be the spacial inhomogeneity of optical field in the plane of p-n junction associated with thermal effects at high pumping level and nonlinearity of gain medium.

As one of possible explanations we can assume that with pumping level increase the new, more complex optical field configurations can take part in laser action. That configurations may have different peak wavelengths and may more effectively consume inversion of carriers population in peripheral regions of pumped stripe contact. These different field configurations can give a contribution to total spectral envelope finally increasing the uniformity of optical field in the plane of p-n junction.

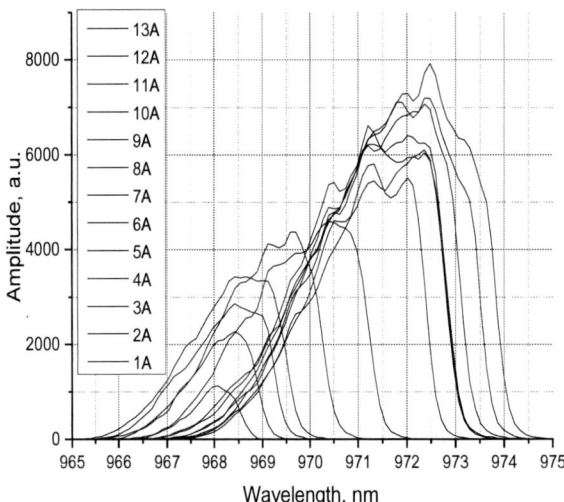

Figure 2. The spectra of laser diode emission at pumping current from 1 to 13 A.

The observed changes in spectral envelope with the increase of pumping current first appear at short wavelength side of the envelope and at higher currents are located at maximum of spectral envelope. This can be the evidence of laser action on other, more complex optical field configurations which propagate in laser resonator at different angles regarding to optical axe, so having different wavelengths as peaks on spectral envelope. The dependence of the peak wavelength of the spectral envelope against the pumping current is presented at figure 3.

At figure 3 we can resolve three typical region of pumping current in which the wavelength dependence against current can be approximated by near to linear with different slope. From threshold current up to pumping current 4 A the slope is around 0.22 nm/A, in the range 4-7 A the slope is 0.85 nm/A and in the range 7-10 A it is 0.26 nm/A. It's interesting that some correlation can be obtained according to the results of spectral envelopes presented at figure 2. Really, in first current region 1-4 A spectral slope at figure 2 has one maximum. At second region 4-7 A we can see the maximum slope of wavelength dependence against current and at the same time in this current range we observe the maximum changes of spectral envelope and formation of configuration with two and three peaks. At third region 7-10 A the slope of wavelength against current dependence is again much less comparably with the second region and close to value for the first region. This third current region corresponds to quasi-stable configuration of the spectral envelope with 3 peaks according to figure 2.

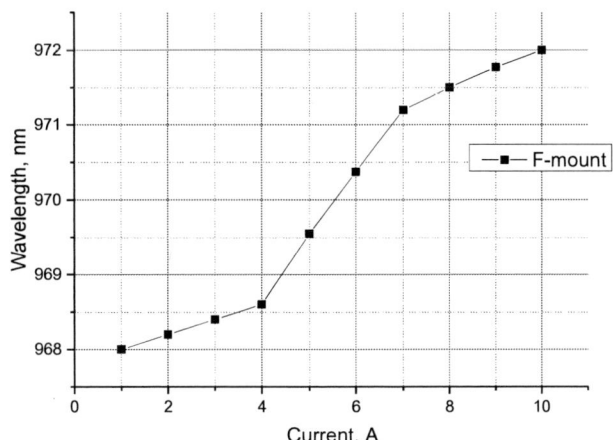

Figure 3. The dependence of the spectral envelope peak against the pumping current.

At figure 4 we present the dependence of spectral envelope half -width against the pumping current.

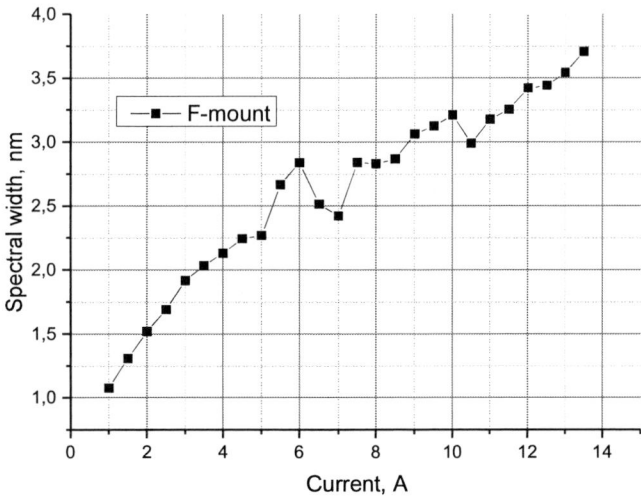

Figure 4. The dependence of spectral envelope half -width against the pumping current.

At this graph we can also conventionally see three regions of pumping current 1-5 A, 5-10 A and 10-14 A. We can also find some correlation with graphs at figure 2 and figure 3. The most smooth is the region from threshold current up to 5 A, and some signs of a slowdown at currents from 3 to 5 A are obvious which can be the evidence of some optical field transformation inside the stripe contact width. For current region from 5 to 10.5 A we see the most serious changes in spectral width which can be the evidence of non-stable optical field distribution. And the last region is more smooth than the second region but less smooth than the first region. This in principle correlates with the spectral envelopes at figure 2. At currents more than 10 A after formation of 3 peaks the spectra with 3 peaks configuration can say becomes quasi-stable.

4. Conclusions

We investigated the optical output power, total efficiency and spectra of two samples of laser diodes assembled in our laboratory on F-mounts with stripe width 95 μm and resonator length 4 mm. Reliable operation with output power 15.1 W under CW conditions was confirmed by testing during 120 hours. Total efficiency in maximum was 72 %, average slope efficiency of W/A characteristic was 1.12 W/A in current range up to 14 A. Some peculiarities in spectral envelope under high current pumping were observed and possible explanation discussed in correlation with the dependences of spectral envelope maximum and its half width against pumping current. More detailed information can be obtained from the results of far field and near field measurements at wide range of pumping current on uniform lot of samples to clear the matter and get reliable knowledge for practical use of such high-power laser diodes at extremely high power and pumping level around 10 -20 times more than the threshold current.

Acknowledgements

This work was carried at National Research Nuclear University MEPhI and supported by the Russian Federation Ministry of Education and Science, Agreement No. 14.575.21.0047, unique identifier of applied scientific research (project) RFMEFI57514X047.

References

[1] Crump P *et al* 2012 *Proc. of SPIE* **7198** 719814-1
[2] Sin Yongkun, LaLumondiere Stephen D, Presser Nathan, Foran Brendan J, Ives Neil A Lotshaw William T and Moss Steven C 2012 *Proc. of SPIE* **8241** 824116-1
[3] Vinokurov D A *et al* 2006 *Fiz. Tekh. Poluprovodn.* **40**(6) 764
[4] Bezotosnyi V V, Krokhin O N, Oleshchenko V A, Pevtsov V F, Popov Yu M and Cheshev E A 2015 *Quantum Electronics* **45**(12) 1088–90
[5] Bezotosnyi V V, Krokhin O N, Oleshchenko V A, Pevtsov V F, Popov Yu M and Cheshev E A 2014 *Quantum Electronics* **44**(2) 145–8
[6] Bezotosnyi V V, Krokhin O N, Oleshchenko V A, Pevtsov V F, Popov Yu M and Cheshev E A 2014 *Quantum Electronics* **44**(10) 899–902

The study of the laser characteristics based on solid solution $Pb_{1-x}Sn_xSe$ ($x \approx 0.07$) emitting at spectral range of 16 μkm

K R Umbetalieva[1,2], K V Marem'yanin[3], V I Gavrilenko[3], I I Zasavitskij[1,2], R R Bitskiy[1] and E A Komochkina[1]

[1]National Research Nuclear University MEPhI, 115409, Moscow, Russia
[2] P. N. Lebedev Physical Institute, Russian Academy of Sciences, 119991, Moscow, Russia
[3] Institute for Physics of Microstructures, Russian Academy of Sciences, 603087, Nizhnii Novgorod, Russia.

E-mail: k.umbetalieva@gmail.com

Abstract. Injection lasers were developed on the basis of the solid solution $Pb_{1-x}Sn_xSe$ for the spectral range of 16 μkm, where the rotational-vibrational absorption lines of heavy molecules are placed. For this purpose, were used, composition the solid solution $Pb_{1-x}Sn_xSe$ near the inversion point $x_i = 0.12$. Single crystals $Pb_{1-x}Sn_xSe$ ($E_g = 0.077$ eV) were grown by directional crystallization from the vapor phase.

1. Introduction
Injection lasers based on A^4B^6-type ternary compounds can be considered quite promising. Semiconductor lasers for the infrared region of the spectrum are useful for some applications (high-resolution molecular spectroscopy, high-sensitivity spectral gas analysis, medical diagnostics [1,2]. Of particular interest is the spectral region of more than 10 μkm, where the rotational-vibrational absorption lines of heavy molecules are placed.

The aim of the research is to study the threshold and the spectral characteristics of Injection lasers based on solid solution $Pb_{1-x}Sn_xSe$ for the 16 μkm spectral region.

2. Band structure $Pb_{1-x}Sn_xSe$
$Pb_{1-x}Sn_xSe$ compound has a NaCl- type crystal structure for a range of compositions $0 \leq x \leq 0,43$. The conduction and valence bands are the mirror image of each other, and their extremes are in the same point of k –space at L-point of Brillouin zone. Figure 1 shows schematically the band structure at 4.2 K. For $Pb_{0.85}Sn_{0.15}Se$ L_6^- and L_6^+ states are degenerate, for $Pb_{0.7}Sn_{0.3}Se$ bands are inverted. The anisotropy coefficient of $Pb_{1-x}Sn_xSe$ compounds is about 2.

Single crystals were grown from the vapor phase [3,4]. P-n junction was produced due to diffusion in the sealed vial from the lead selenide charge enriched either by chalcogen for n-type crystals or by metal for p-type crystals. Resonators were formed by cleaved planes (100).

Content from this work may be used under the terms of the Creative Commons Attribution 3.0 licence. Any further distribution of this work must maintain attribution to the author(s) and the title of the work, journal citation and DOI.
Published under licence by IOP Publishing Ltd

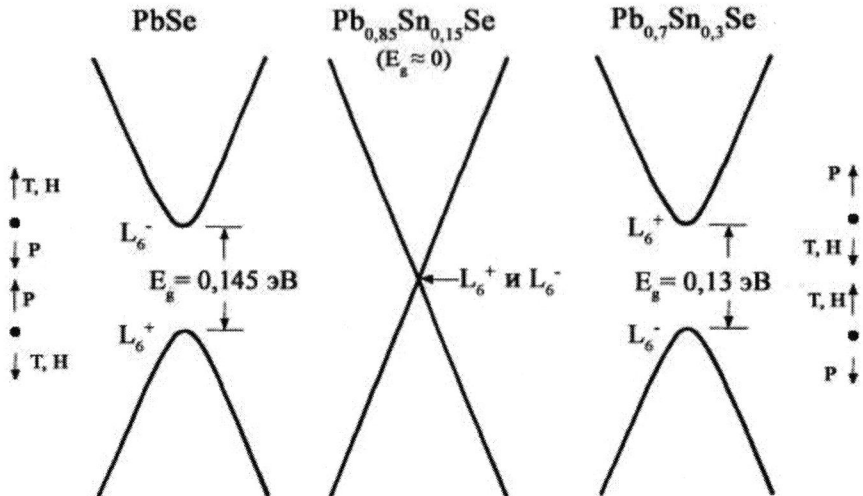

Figure 1. The inversion of the conduction band and valence band in $Pb_{1-x}Sn_xSe$ compounds as a result of the change in composition.

3. Experimental procedure

To measure the integrated intensity and emission spectra of the lasers it is necessary to use the facility shown in figure 2.It works as follows.

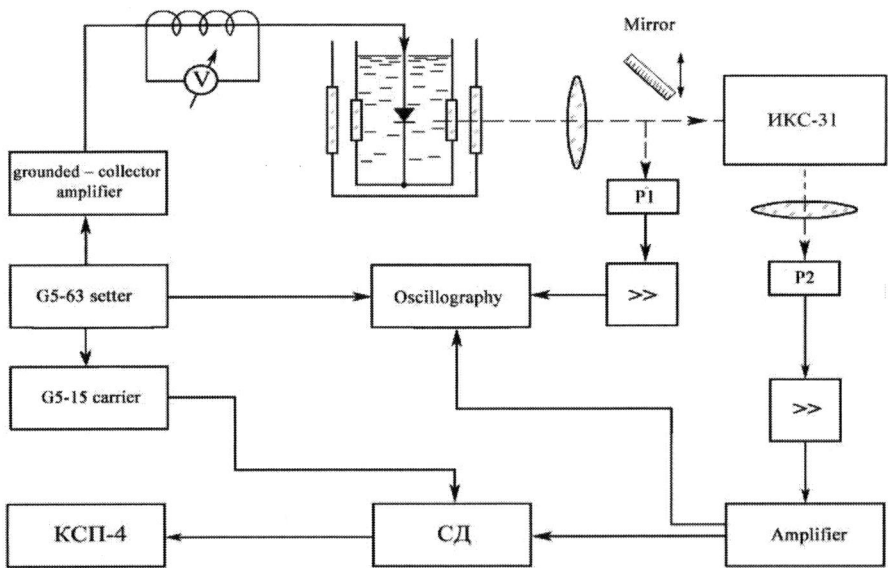

Figure 2. The setup scheme for measuring electroluminescence.

A pulse was fed to the power amplifier (grounded –collector amplifier) from G5-63 setter (oscillator), from which the current pulse was fed to the sample placed in the optical cryostat. Pulse voltmeter measured the output voltage. Diode emission was focused by a lens and then, depending on the measurements, was directed either through the mirror to the receiver R1 based on Ge: Au for measuring the integrated intensity, or to the entrance slit of the X-31-type grating monochromator (IR spectroscopy) (echelette 75 lp / mm.) for removing the emission spectrum.

At low temperature (~ 20 C) emission spectra, measurements were conducted using the vacuum-Fourier spectrometer [4] to eliminate atmospheric absorption.

4. Emission characteristics of lasers $Pb_{1-x}Sn_xSe$

We have experimentally obtained the threshold and spectral emission characteristics of $Pb_{1-x}Sn_xSe$ -based lasers (x≈ 0.07) in the 16 μkm spectral region at low temperatures (20 - 80 K).

4.1. Threshold characteristics of lasers

Threshold current was measured at liquid-helium temperature. Current-voltage characteristics (CVC) is the important characteristics of diode lasers. It determines such laser characteristics as the band gap (E_g) of the material, the threshold current, the series resistance. Since the laser diode operates at a forward bias and a large level of injection, a direct branch of the current-voltage characteristics of lasers was usually measured.

Figure 3 shows the measured values at T = 4.2 K for lasers $Pb_{1-x}Sn_xSe$ (x ≈0.07). It can be seen that the characteristics is a straight line, the slope of which is determined by the series resistance of the laser assembly, which can be calculated as R= (dU/dI) = 0.5 Ohm. Extrapolation of the straight portion to the abscissa characteristics allows defining sufficiently exactly the difference in contact potential of U_k at the p-n junction and thus determining the value E_g.

Dependence of the emission intensity on the current. At the liquid nitrogen temperature the values of the threshold current increase from 5 to 15 A. At low temperatures (T ≤ 20K) the laser threshold current ranges from 2 to 8 A. Therefore, continuous operation was impeded and all studies were carried out in a pulsed mode. Figure 4 shows the dependence of the emission intensity on the current.

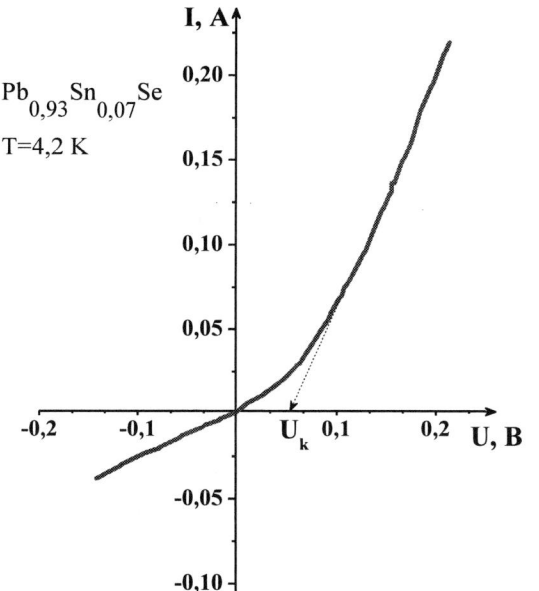
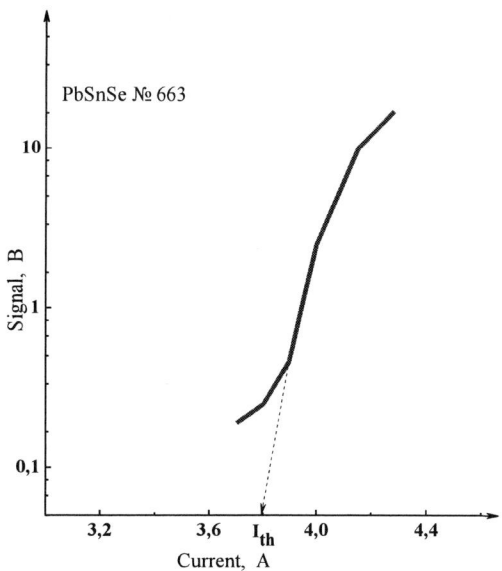

Figure 3. Current-voltage characteristics of $Pb_{1-x}Sn_xSe$ (x ≈0.07) at helium temperature.

Figure 4. Dependence of the emission intensity on the current for $Pb_{1-x}Sn_xSe$ (x≈0.07) at nitrogen temperature.

This watt-ampere dependence has three sections. In the first section (the curve goes up very slightly) – the dependence close to the linear one (the more the current increases, the more the signal changes) - corresponds to spontaneous emission. The second section (the curve goes up sharply) - when a small change in the current significantly changes the signal. This section is related to stimulated emission. Subsequently, luminescence spectra showed that it was not just stimulated emission but oscillating mode (laser effect). Finally, the third section (flatter curve, again) – saturation state - when the laser is heated by the current, and its intensity decreases. The laser heating is associated with the ohmic heating of contacts and non-radiative re-combination of the active region, which also leads to heating.

4.2. Emission spectra of $Pb_{1-x}Sn_xSe$-lasers
The lasers operated at $T = 19$ K in pulsed mode ($T = 1$-5 ms) with the repetition frequency (0.17-1 kHz). When there occurred a small excess over the current threshold, the lasers worked in the single-mode oscillation. When there was a strong current excess in the emission spectrum, a series of equidistant longitudinal modes (figure 5) occurred, with transverse modes superimposed on some of them.

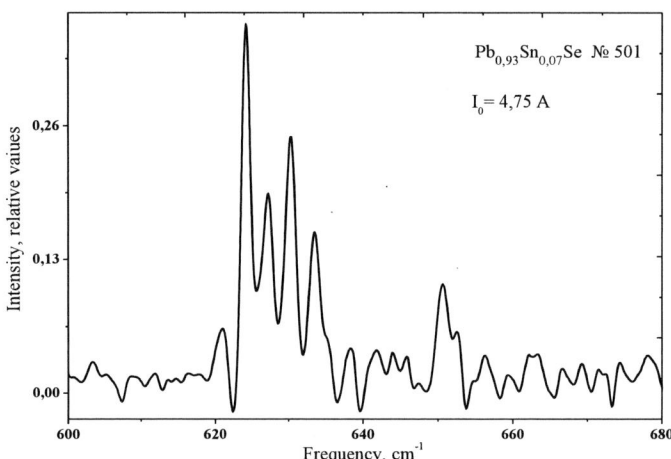

Figure 5. Emission spectrum of $Pb_{0.93}Sn_{0.07}Se$-laser № 501.

In the emission spectrum of $Pb_{0.93}Sn_{0.07}Se$-laser № 501 there are equidistant longitudinal modes. The distance between the longitudinal modes is equal, and it is about 3 cm^{-1}.

Let us consider the shift of the emission frequency of individual modes of lasers $Pb_{0.93}Sn_{0.07}Se$. Emission spectra are measured as the current changes. The rate of emission frequency tuning of 5 individual modes of lasers was calculated according to the measurements made. For this purpose, samples of $Pb_{1-x}Sn_xSe$ № 416 and № 474 with the same composition ($x \approx 0.07$) and the same emission wavelength, but with different lengths of the resonator were selected. Figure 6 shows the dependence of emission frequency on the current. Lasers emit at the wavelength of about 16 μkm. Frequency increases as the current increases. This is due to the fact that the width of the band gap E_g increases with an increase in temperature, and the sample is heated slightly with an increase in the impressed current.

Figure 7 shows the spectra of laser emission depending on the current injection at 19 K. With the current increasing, the stimulated emission of the laser increases through forming different generation modes. At larger currents a greater amount of modes in the emission spectrum is separated. Most lasers operate in the multimode regime. The main reasons for multimode nature are as follows. Each mode is unique in its spatial heterogeneity and stationary localization in the active medium. Sections of the active laser medium, located in the wave nodes, do not almost give off their energy to the lasing mode. The level of inverse population of such sections increases as pumping increases. Therefore, there are favorable conditions for generating other types of waves whose antinodes and nodes are located differently in space.

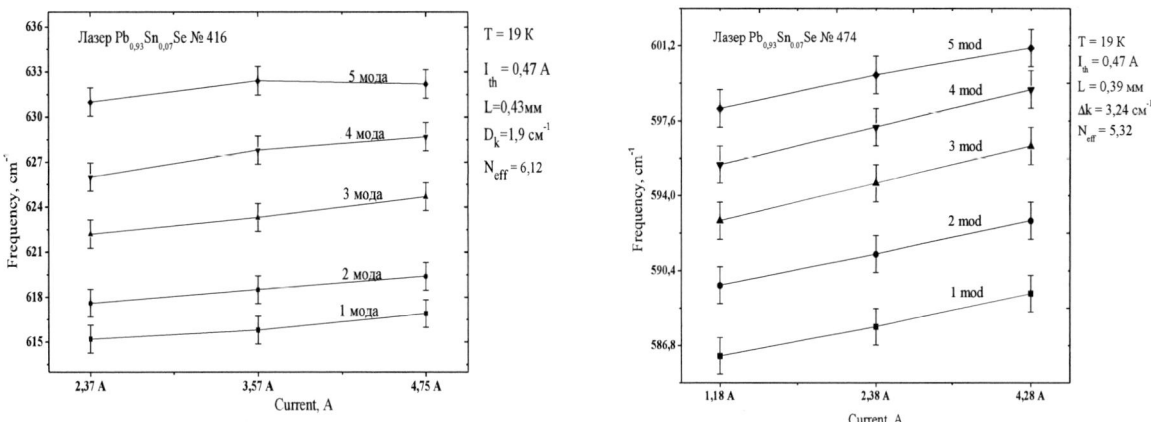

Figure 6. The shift of the emission frequency of individual modes of lasers $Pb_{0.93}Sn_{0.07}Se$ № 416 (left) and №474 (right).

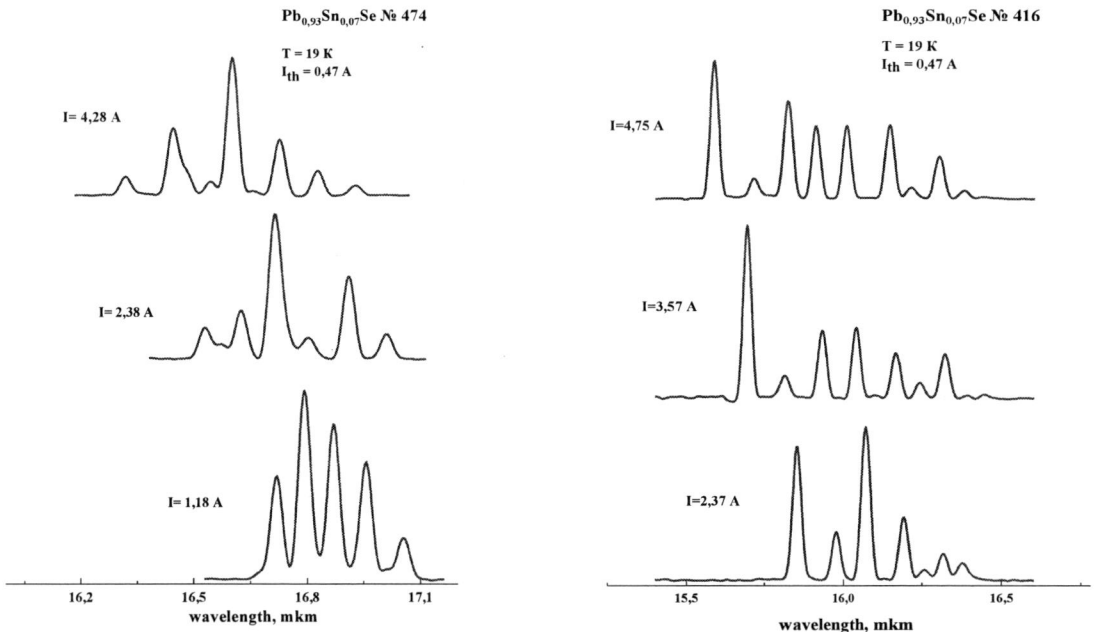

Figure 7. Emission spectra of lasers $Pb_{0.93}Sn_{0.07}Se$ № 416 (left) and № 474 (right).

According to our measurements, the effective refractive index was calculated by the formula:

$$\Delta k = \frac{1}{2LN^*}$$

As seen from the calculation, for a diode laser Pb0.93Sn0.07Se № 501 with a series of equidistant modes, at the measured inter-mode distance $\Delta k = 3.2$ cm^{-1} and resonator length $L = 0.25$ mm the effective refractive index is $N^* = 6.25$ for a given composition of the solid solution $x \approx 0.07$. With account for the data from [1,2], the dependence of the effective refractive index on the composition (figure 8) $N^* = 5.1 + 17x$ was obtained.

Table 1. Effective refractive index of the active medium

x	Laser number	L, mm	T = 19 K				
			I_{th}, A	Δk, cm^{-1}	N*	Tuning	
						10^{-3}, cm^{-1}/ mA	MHz / mA
0.07	416	0.43	0.47	1.9	6.1	1.4	40
0.07	474	0.29	0.47	3.24	5,32	1.04	30
0.07	501	0.25	2.38	3.2	6.25		

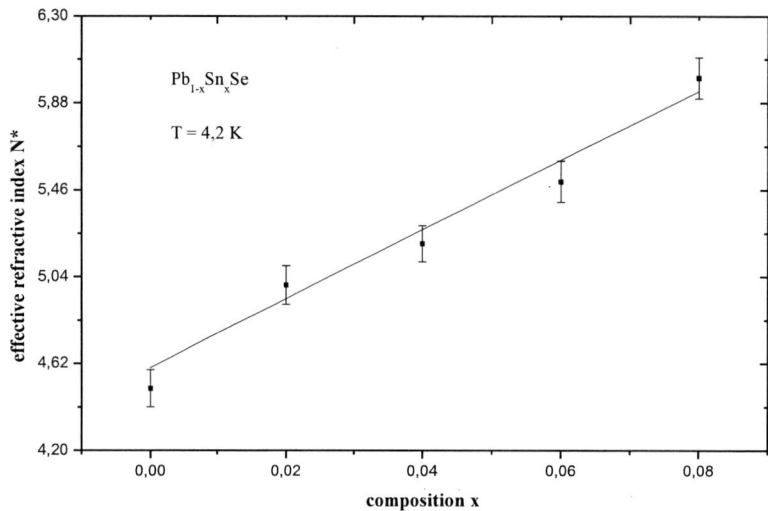

Figure 8. Dependence of the effective refractive index on composition.

5. Conclusion

1. Diode lasers for the spectral region of about 16 μkm have been developed on the basis of the solid solution $Pb_{1-x}Sn_xSe$.

2. Contact potential difference and resistance have been measured by means of the current-voltage characteristics.

3. The threshold current density of diode lasers has been defined from watt-ampere characteristics over the extrapolation of the steep section to the horizontal axis.

4. The working temperature of the laser in the pulse mode was in the range from 4.2 to 77.4 K. Threshold currents varied from 0.5 to 2 A, respectively.

5. The effective refractive index has been determined from the emission spectra, depending on the composition and tuning rate of individual modes with the current.

Acknowledgments

This work was supported by Competitiveness Program of National Research Nuclear University MEPhI.

References

[1] Dornhaus R, Nimtz G, Schlicht B 1983 *Narrow-Gap Semiconductor* (Berlin: Springer-Verlag) 309

[2] Zasavickij I 1993 *Proc. of Physical Institute of the Russian Academy of Sciences* Opt. and electr. prop. of semiconductors **224** 3
[3] Tacke M 1995 *Infrared Phys. and Technol.* **36** 447
[4] Maremyanin K, Ikonnikov A, Antonov A, Rumyantsev V, Morozov S, Bovkun L, Umbetalieva K, Chizhevskiy E, Zasavitskiy I and Gavrilenko V 2015 *Semiconductors* **49**(12) 1672-6

Middle infrared Fe^{2+}:ZnS, Fe^{2+}:ZnSe and Cr^{2+}:CdSe lasers: new results

V I Kozlovsky[1,2], Y V Korostelin[1], Y P Podmar'kov[1,3], Y K Skasyrsky[1] and M P Frolov[1,3]

[1]P N Lebedev Physical Institute, Leninsky pr. 53, 119991, Moscow, Russia
[2]National Research Nuclear University MEPhI (Moscow Engineering Physics Institute), 31 Kashirskoye shosse, 115409, Moscow, Russia
[3]Moscow Institute of Physics and Technology (State University), 9 Institutskii per., 141700, Dolgoprudny, Moscow region, Russia

E-mail: vikoz@sci.lebedev.ru

Abstract. The output energy of 10.6 and 3.25 J with a pulse duration of ~1 ms and a crystal temperature of 85 K was achieved in the Fe:ZnSe and Fe:ZnS lasers, respectively. At room temperature, the Fe:ZnSe laser energy was as high as 1.2 J with a pulse duration of 150 ns. A few optical schemas of the Cr:CdSe laser with laser diode pumping were realized.

1. Introduction

Lasers of middle infrared (mid IR) spectral range are perspective for various applications in laser location of objects in atmosphere, high resolution spectroscopy, metrology, monitoring of pollution of the environment, medicine. To date the works directed to increasing an average power up to 10 W and pulse energy up to 1 J or higher in the 4-5 μm range are of the greatest interest.

Among well-known lasers of this spectral range it is worth to mark out chemical DF lasers (toxic, bulky, ineffective at $\lambda > 4$ μm), quantum cascade lasers (high level technology, problem with obtaining high energy pulse) and optical parametric oscillators (unreliable as yet). In recent years, the lasers based on II-VI compound crystals doped Cr and Fe, which are free from disadvantages of mentioned lasers, are intensively developed.

To date the following results are achieved. At previous symposium, we reported on obtaining of 2.1 J in the pulsed Fe^{2+}:ZnSe laser at liquid nitrogen cooling of the active crystal pumped by an Er:YAG laser [1]. An average power of such a laser operating in a repetitively pulsed mode with cryogenic cooling of the crystal was increased up to 35 W [2]. The maximum energy of 253 mJ was obtained at room temperature of the crystal pumped by a HF laser [3].

In this talk, we report new results obtained by our team and other research groups using ZnSe:Fe crystals grown by our team in P.N. Lebedev Physical Institute. Also we present several optical schemas of Cr^{2+}:CdSe lasers pumped by laser diodes. We hope that the Cr:CdSe laser may be used as a pump for the Fe:ZnSe laser in future.

2. High energy pulsed Fe:ZnSe and Fe:ZnS lasers at pumping by Er:YAG laser radiation

---

Content from this work may be used under the terms of the Creative Commons Attribution 3.0 licence. Any further distribution of this work must maintain attribution to the author(s) and the title of the work, journal citation and DOI.
Published under licence by IOP Publishing Ltd

Figure 1 presents the schema of the experimental setup for studying Fe:ZnS and Fe:ZnSe lasers upon pumping by a free-running flashlamp-pumped Er:YAG laser. This schema is described in detail in [1, 4].

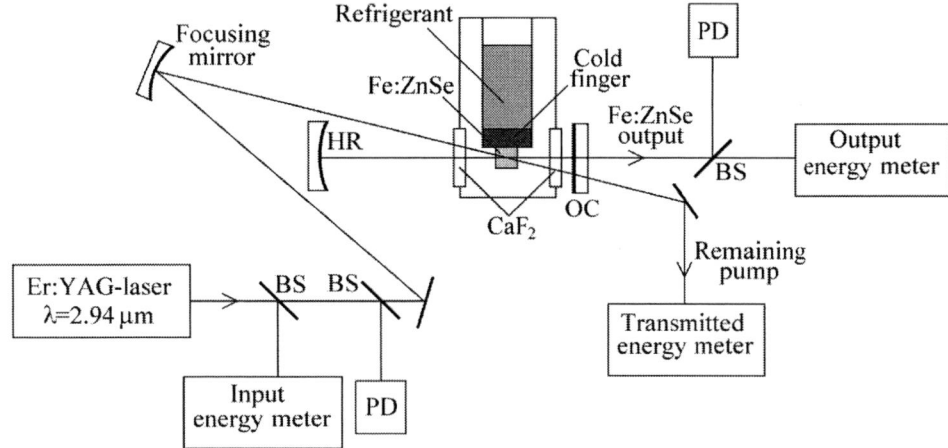

Figure 1. Schema of the experimental setup for studying Fe:ZnSe and Fe:ZnS lasers. PD is photodiode, BS is beam splitter, HR is high reflective mirror, OC is output coupler.

The main results are presented in figure 2a. The most experiments were fulfilled at the pump pulse

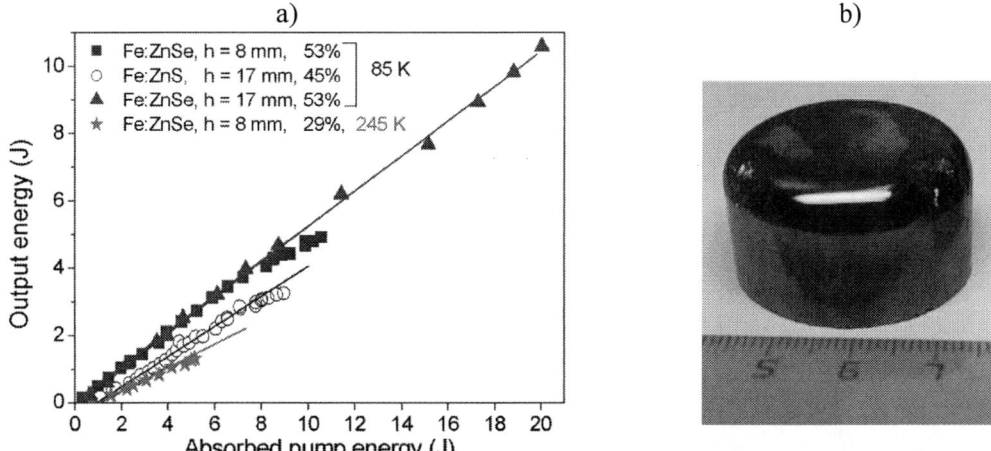

Figure 2. Dependences of the output energy of the pulsed Fe:ZnSe/ZnS laser on the absorbed pump energy at using of different active elements (a) and a ZnSe:Fe boule used in the last experiments with an active element length (h) of 17 mm (b). Crystal temperature and slope efficiency are indicated in figure 2a also.

duration of 0.9 ms and the pump energy of 15 J [4]. The maximum output energy of the Fe:ZnSe laser was 4.9 J at the crystal temperature of 85 K and 1.3 J at T = 245 K. The last temperature may be controlled by a two-step thermoelectric cooler. The energy of 3.25 J was achieved with the ZnS:Fe crystal. Recently we were able to increase the pump energy up to 30 J at a little longer pump pulse duration of 1.3 ms. However we succeeded in increasing pulse energy of the Fe:ZnSe laser only by using a larger active crystal with smaller Fe concentration (10^{18} cm^{-3}). At that case, a problem of inversion dumping by amplified spontaneous noise at large lateral size of pump beam spot (up to 16 mm in diameter in our last experiments) was solved. The active element was fabricated from a boule presented in figure 2b. The maximum energy of the Fe:ZnSe laser was as high as 10.6 J at $T = 85$ K while the slope efficiency was 53% relative to absorbed input energy. In the near future, we hope to obtain a pulse energy of 5 J or higher with a thermoelectric cooler.

3. Fe:ZnSe laser with pumping by HF laser radiation

Pulse-periodic operation mode of the Fe:ZnSe laser was studied at pumping by an electric discharge HF laser in the work [5]. The main results of this study are presented in figures 3 and 4. The HF laser

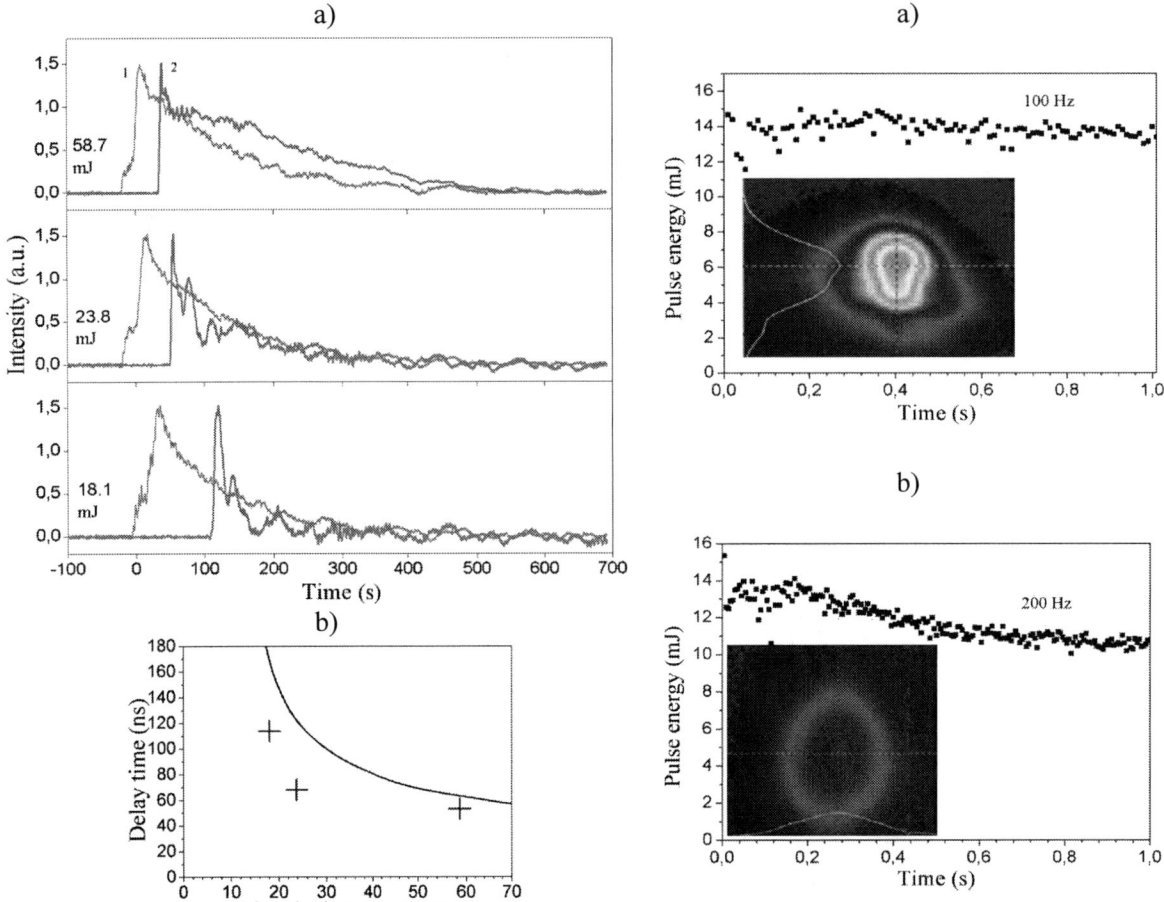

Figure 3. Oscillograms of the pulses of the HF pump laser (1) and Fe:ZnSe laser (2) at pump pulse energies of 18.1, 23.8, and 58.7 mJ (a), as well as dependence of the laser pulse delay with respect to the pump pulse at a level of 0.1 of the intensity on the absorbed pump energy (b) (the points and the curve correspond to experiment and calculation, respectively).

Figure 4. Changes in the laser pulse energy (a, b) from pulse to pulse (each point is an individual pulse) in 1-s trains with different pulse repetition rates 100 (a) and 200 (b) Hz. Near- (insert a) and far-field (insert b) patterns of the Fe:ZnSe laser radiation

operated in the pulse periodic regime with a repetition rate of 100 and 200 Hz for a train with duration of 1 s. The crystal was at room temperature. Each pulse in the train had duration of about 100 ns at half maximum (see figure 3a). The pump energy was as high as 75 mJ. The laser pulse contained relaxation spikes and was delayed with respect to the pump pulse. This delay decreased with increasing pump energy (figure 3b).

The energy of an individual pulse was about 14 mJ at an absorbed energy of 58 mJ and weak changed from pulse to pulse in the train at the repetition rate 100 Hz (figure 4a). At 200 Hz, a small drop of the output energy of an individual pulse was observed because of crystal heating (figure 4b). The average train power was 1.4 and 2.4 W at pulse repetition rates of 100 and 200 Hz, respectively. The total train energies were, respectively, 1.4 and 2.4 J. Lateral distribution in the near-field pattern was elliptic that was responsive to a form of pump spot with the average lateral size of 2.2 mm. The total divergence angle at a level of 1/e of the maximum intensity was 3.8 mrad.

To achieve high pulse energy at room temperature of the crystal, the high energy pulsed HF laser was used in work [6]. The maximum energy was obtained as high as 1.2 J at the input pulse energy of 4.8 J and the pulse duration of 150 ns (at half maximum).

4. Amplifier of nanosecond pulses based on ZnSe:Fe crystals

First experiments concerning amplification of nanosecond pulses in the ZnSe:Fe crystals were carried out. In future we plan to develop this work to obtain high power femtosecond pulses [7]. A schema of an experimental setup consisting of a master oscillator and an amplifier is presented in figure 5a.

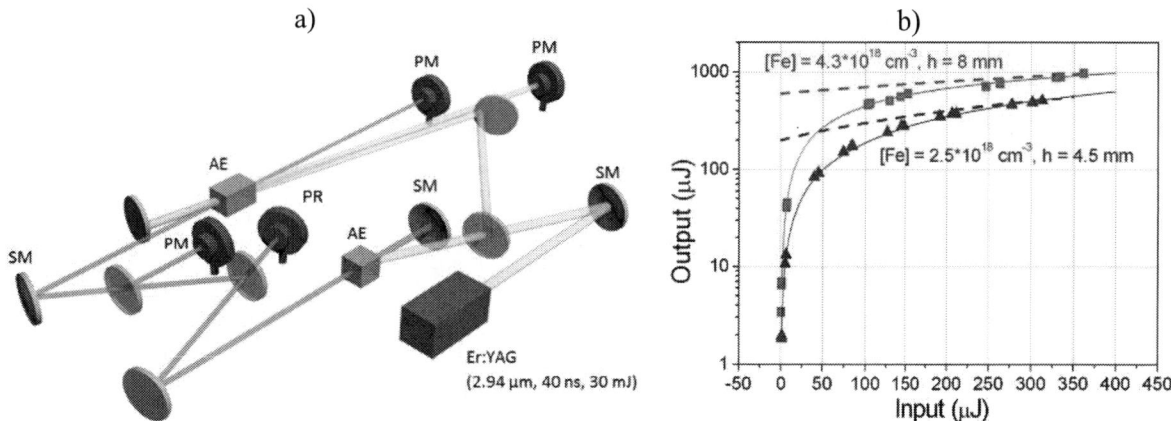

Figure 5. Schema of a setup for studying amplification of nanosecond pulses in the ZnSe:Fe crystal (a) and output – input characteristics of the amplifier for two different ZnSe:Fe crystals (b). AE is active element, PM is power meter, PR is photoresistance, SM is spherical mirror.

The passively Q-switched Er:YAG flashlamp laser was used by us to pump the ZnSe:Fe crystals. It generates laser pulses with duration of 40 ns and energy of 30 mJ. The incident pump beam was divided by a beam splitter into two beams that were focused by a spherical mirror (radius of curvature is 150 cm) in the ZnSe:Fe crystals which were placed in both the oscillator and the amplifier. The pump energy directed to the oscillator and amplifier were 7 mJ and 23 mJ, respectively.

The master oscillator cavity was formed by a spherical mirror (radius of curvature is 50 cm) and a well-polished end face of the ZnSe:Fe crystal (without any antireflective coating). The laser was operated at room temperature with a central wavelength of 4.4 μm. The seed pulse from the oscillator was focused into the ZnSe:Fe crystal using the spherical mirror with a curvature radius of 100 cm. The FWHM pulse duration was 15 ns, and the energy was up to 360 μJ.

To obtain gain parameters we tested two ZnSe:Fe crystals with antireflection coating on a central wavelength of 4.4 μm and the different doping of divalent iron ions in a ZnSe matrix. The samples were cut into rectangular slabs of 5×5 mm^2 cross sections, the first one with a thickness of 4.5 mm and 2.5×10^{18} cm^{-3} doping, and the second one with a thickness of 8 mm and 4.2×10^{18} cm^{-3} doping. The diameters of the seed beam and pump beam were about 1.2 mm and 2 mm, respectively. The remaining part of the pump was returned into the active medium.

Comparison of dependences of the output pulse energy on the seed pulse energy for two crystals was presented in figure 5b. The maximal 2.7 times amplified pulse was about 1 mJ for the crystal with Fe concentration of 4.3×10^{18} cm^{-3}.

5. Cr:CdSe laser with pumping by laser diodes

High energy characteristics of the Fe:ZnSe laser described above was achieved using the bulky HF laser and flashlamp Er:YAG laser. For practical uses, a laser version with pumping by solid-state laser radiation, which in turn is pumped by laser diodes, is preferable one. It is desirable also that the all elements of a laser system operate at room temperature. At that case, the ZnSe:Fe crystal should be pumped by a short pulse train. There are some problems to create a repetitively pulsed Er-laser ($\lambda \approx 3$ μm) with nanosecond pulse duration and pumping by laser diodes. But several successful versions of a similar laser based on Tm ions were already developed. Unfortunately Tm-lasers emit at a wavelength close to 1.9 μm and cannot directly be used as a pump for the Fe:ZnSe laser. We offered to convert a radiation with wavelength of 1.9 μm to 2.9-3 μm using a laser based on a CdSe:Cr crystal [8]. A Tm-fibre laser was used for pumping of the Cr:CdSe laser. Recently we were able to start up the Cr:CdSe laser directly pumped by a fibre-coupled bar of laser diodes with radiation wavelength of 1.94 μm [9]. However both these schemas do not allow to create high power nanosecond pulses for pumping of the Fe:ZnSe laser at room temperature.

High enough power pulses may be obtained by a laser-diode pumped solid-state Tm-laser. One of possible schemas of the Cr:CdSe laser pumped by a Tm:YAP laser radiation ($\lambda = 1.99$ μm) realized in a work [10] with our crystals is presented in figure 6a. A Fabry-Perot etalon and an interference-

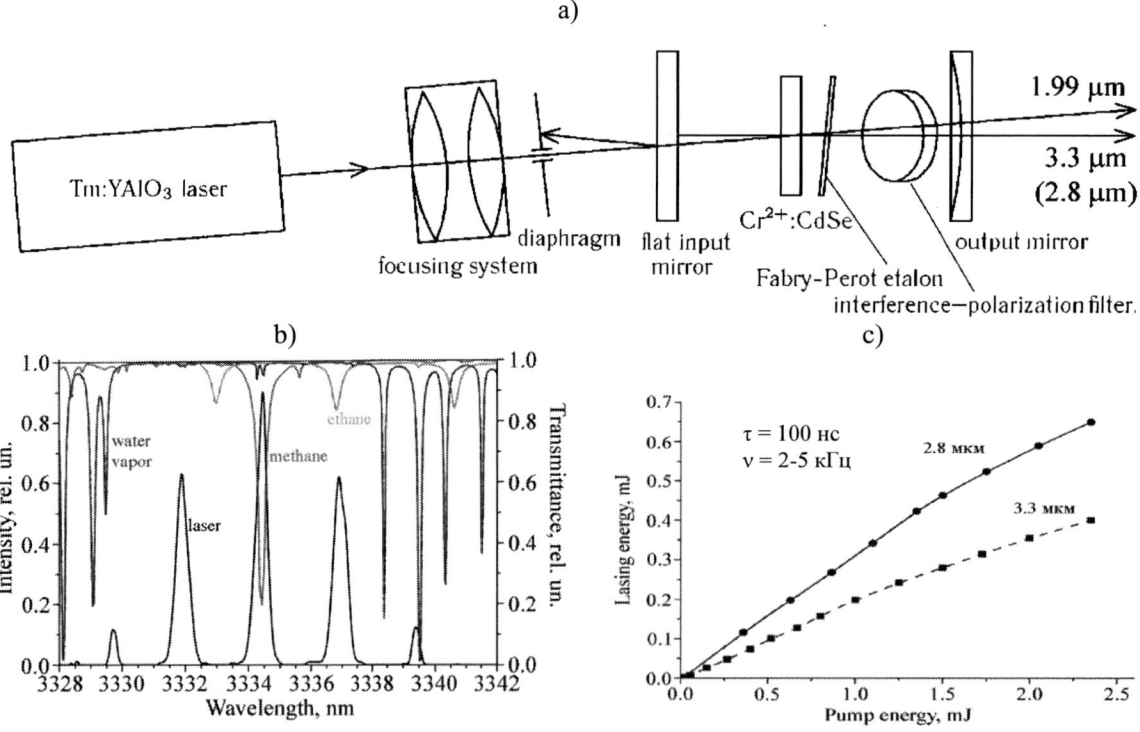

Figure 6. Schematic of the experimental setup (a), spectrum of laser radiation and transmission spectra of methane, ethane and water vapors (b) and dependences of the laser pulse energy on the pump energy at two values of laser wavelength: 2.8 and 3.3 μm (c).

polarization filter were placed inside a cavity to control spectrum near wavelength of 2.8 μm or 3.3

μm. An example of such a laser spectrum is presented in figure 6b. Here the spectra of absorption by methane, ethane and water vapours are shown also. It is seen that we can turn the laser spectrum such a manner that two of the most intense laser lines will be matched with the ethane and methane absorption lines, 3336.8 and 3334.3 nm, respectively. While the third intense line at 3331.8 nm not coinciding with any lines of ethane, methane or water may be taken as a reference line. Such a laser is of independent interest of application as a lidar.

Figure 6c presents dependences of the pulse energy of the Cr:CdSe laser on the pump energy at two values of the laser wavelength. At pulse duration of 100 ns and repetition rate of 2-5 kHz, the maximum energy was 0.65 and 0.4 mJ at wavelength of 2.8 and 3.3 μm, respectively.

For pumping of the Cr:CdSe laser was also used a repetitively pulsed $Tm^{3+}:Lu_2O_3$ ceramic laser with radiation at a wavelength of 2.066 μm [11]. At that case, lasing was realized at a wavelength of 2.92 μm with the parameters of output pulse energy, pulse duration and repetition rate being close to ones described above. We plan to continue this work to increase an energy in individual pulses of the Cr:CdSe laser and use a pulsed train for pumping the Fe:ZnSe laser.

Acknowledgements
This work was supported by Competitiveness Program of National Research Nuclear University MEPhI.

References
[1] Frolov M P, Korostelin Yu V, Kozlovsky V I, Mislavskii V V, Podmar'kov Yu P, Savinova S A and Skasyrsky Ya K 2013 *Laser Physics Letters* **10** 125001
[2] Mirov S, Fedorov V, Martyshkin D, Moskalev I, Mirov M and Vasilyev S 2015 *Proc. of SPIE* **9467** 94672K
[3] Gavrishchuk E M, Ikonnikov V B, Kazantsev S Yu, Kononov I G, Rodin S A, Savin D V, Timofeeva N A and Firsov K N 2015 *Quantum Electron* **45** 823
[4] Frolov M P, Korostelin Yu V, Kozlovsky V I, Podmar'kov Yu P, Savinova S A and Skasyrsky Ya K 2015 *Laser Physics Letters* **12** 055001
[5] Velikanov S D *et al* 2015 *Quantum Electron* **45** 1
[6] Firsov K N *et al* 2016 *Laser Physics Letters* **13** 015002
[7] Potemkin F V *et al* 2016 *Laser Physics Letters* **13** 015401
[8] Voronov A A, Kozlovsky V I, Korostelin Yu V, Landman A I, Podmar'kov Yu P, Skasyrskii Ya K and Frolov M P 2008 *Quantum Electron* **38** 1113
[9] Lazarev V A, Tarabrin M K, Kovtun A A, Karasik V E, Kireev A N, Kozlovsky V I, Korostelin Yu V, Podmar'kov Yu P, Frolov M P and Gubin M A 2015 *Laser Physics Letters* **12** 125003
[10] Zakharov N G *et al* 2015 *Bulletin of the Lebedev Physics Institute* **42** 216
[11] Antipov O L, Eranov I D, Frolov M P, Korostelin Yu V, Kozlovsky V I, Novikov A A, Podmar'kov Yu P and Skasyrsky Ya K 2015 *Laser Physics Letters* **12** 045801

Empirical mode with a variable spatial-temporal structure and the dynamics of superradiant lasers

E R Kocharovskaya[1], A S Gavrilov[1], V V Kocharovsky[1,2],

E M Loskutov[1], D N Mukhin[1] A M Feigin[1] and Vl V Kocharovsky[1]

[1]Institute of Applied Physics of the RAS, 603950, Nizhny Novgorod, Russia
[2]Department of Physics and Astronomy, Texas A&M University, College Station, TX, USA

E-mail: kochar@appl.sci-nnov.ru

Abstract. An approach of the empirical modes with a variable spatial-temporal structure is proposed and developed for the analysis of non-stationary nonlinear dynamics of the multimode superradiant lasers with a low-Q cavity and a strong inhomogeneous broadening of lasing transition in an active medium. It is shown that the approach makes it possible to analyze a number of complicated dynamical phenomena in an ensemble of the strongly interacting centers which constitute the active medium and are exposed to CW pumping.

1. Introduction

A non-stationary lasing under CW pumping is a typical dynamical regime which takes place well above a lasing threshold. This regime, as a rule, shows the multimode oscillations which caused by nonlinearity of an active medium, but also may be originated from and complicated by various additional elements inserted into a laser, e.g., by nonlinear absorbers, optical delay feedback, etc. [1-4]. A multimode lasing may exhibit highly non-trivial and even chaotic oscillations which are used in an optical information processing and a wideband spectroscopy. Usually, however, such oscillations are caused by an interaction of cavity modes which are quasi-stationary (weakly modulated) modes and have well-defined spatial structures dictated by an amplification in active medium and by the cavity features, e.g., reflections from the facet mirrors or distributed feedback reflections. Under these conditions, a spatial-temporal evolution of a laser field is described by a superposition of the cavity modes with the time-dependent amplitudes and the spatial profiles, which are fixed and known beforehand. For the standard lasers with high-Q cavities, a resulting dynamical spectrum of the field is usually quasi-equidistant over frequency and homogeneous in time. For a one-dimensional model studied in the present paper, the high-Q cavity modes have the field envelops which are almost constant along a path of wave propagation.

A laser dynamics becomes essentially different in the case of lasers with *low-Q (bad) cavities* where a photon lifetime, T_E, is less than a polarization relaxation time (a lifetime of the optical dipole oscillations), T_2, of the individual active centers excited by pumping. For simplicity's sake, we refer to a well-known two-level model of active medium [1, 4-7] and, for definiteness, we consider the case of a strong inhomogeneous broadening of the lasing transition, $2/T_2^* >> 2/T_2$. Actually, a theoretical analysis of the spatial-temporal dynamics of the field and its spectral and correlation features in such lasers, known as the superradiant lasers, cannot be based on the above-mentioned standard decomposition on neither 'cold' no 'hot' modes which are defined, respectively, by a cavity without or with taking into account the active medium (under the condition of a steady-state homogeneous population inversion of the lasing energy levels). The point is that the polarization of active centers in the superradiant lasers does not follow adiabatically a value of the local electric field and plays a part of an independent dynamical variable. The field itself flows out of a cavity rapidly and changes in time and space strongly. These effects result in a complicated spatial-temporal dynamics of the population inversion and, hence, lead to an efficient non-adiabatic coupling of cold and/or hot modes what makes those modes useless for the interpretation of collective emission phenomena, even in the case of a steady time-dependent lasing under CW pumping.

A progress in modern technologies, especially in the field of semiconductor heterostructures, leaves no doubts about near fabrication of the dense (in space and spectrum) ensembles of active centers needed for such dynamically rich lasing in the low-Q cavities under CW pumping. (In the case of a pulsed pumping, this kind of a single-burst radiative cooperative phenomenon, known as a collective spontaneous emission, or superfluorescence, has been observed and verified in a number of active-center ensembles, e.g., the ensembles of quantum dots, impurity centers, excitons, and free electrons and holes in quantum wells placed in a quantizing magnetic field [7-15].) Among various regimes, the superradiant lasing includes, first of all, a generation of a sequence of the coherent bunches of pulses of collective spontaneous emission (Dicke superradiance) which is possible at a high rate of pumping, i.e., at a short enough time of incoherent relaxation and creation of inversion of the lasing energy levels of active centers, T_1. The related dynamics of inversion may also provide the conditions for a partial self-locking of the quasi-stationary laser modes which would result in another sequence of pulses with a repetition rate defined by the cavity round-trip time. The expected pulse durations in both sequences lie in the picosecond and/or subpicosecond timescale that gives good prospects for the superradiant-lasing applications in the optical information technologies and the diagnostic techniques for fundamental physics of many-particle systems.

In order to understand the time-dependent space-inhomogeneous configurations of a lasing field in various steady-superradiant regimes and to interpret them properly, we suggest to use a well-known approach of empirical orthogonal functions (EOFs) which is based on a method of the main components [16] and have been widely employed in the analysis of observed space-distributed time series [17-19], including the correlation analysis of data [16, 20]. For the laser problem under consideration, this approach should be generalized and then may be called as an approach of *Space-Time Empirical Modes* (STEMs). The definition and applications of the newly suggested modes are illustrated below on the basis of a numerical solution to the integral-differential Maxwell-Bloch equations for a 1D model of a cavity and an active medium with strong inhomogeneous broadening of a spectral line, $2/T_2^* \gg 2/T_2$. Namely, we consider a low-Q hybrid Fabry-Perot cavity with a distribute feedback (DFB) of the counter-propagating waves, where a generation of the superradiant pulses may be accompanied by a partial self-locking of the longitudinal modes without use of any additional technique of mode locking [21, 22].

2. Model of a superradiant laser

According to a preliminary qualitative analysis of dynamics of a superradiant laser with high spatial and spectral density of active centers [23, 24], an output radiation under CW pumping consists, as a rule, of one or several quasi-chaotic sequences of ultrashort powerful pulses and cannot be described as a superposition of any hot modes calculated under the condition of a given homogeneous inversion of a lasing transition in active medium. For a hybrid DFB – Fabry-Perot cavity, a hot mode contains two symmetric counter-propagating waves, each being a sum of two inhomogeneous spatial harmonics with the close wave numbers slightly shifted from the Bragg (DFB) resonance. The hot-mode consideration is sufficient only for the analysis of a quasi-stationary generation regimes with relatively weak and long-term modulation of the mode amplitudes or for the evaluation of a lasing threshold as a condition of the hot-mode instability [1, 5-7, 22], $v_c^2 T_2^* T_E > 1$, where $v_c = \sqrt{2\pi d^2 N_0 \omega_{21} / \hbar \bar{\varepsilon}}$ is a so-called cooperative frequency of an ensemble of active centers with a density N_0, d a dipole moment of an active center at the lasing-transition frequency ω_{21}, $\bar{\varepsilon}$ an average dielectric permittivity of an active medium. In our case of the strong inhomogeneous broadening of spectral line, when a parameter $\Delta_0 = \left(v_c T_2^*\right)^{-1}$ is large, $\Delta_0 \gg 1$, the lasing threshold is defined by a so-called active cooperative frequency [15, 22, 23]: $\bar{V}_c \equiv v_c / \Delta_0 > 1 / T_E$.

In the considered 1D two-level laser model, the dynamics of the field inside a cavity,

$$\text{E=Re}\left[\left(\tilde{a}_+(z,t)e^{ik_0 z} + \tilde{a}_-(z,t)e^{-ik_0 z}\right)e^{-i\omega_0 t}\right], \tag{1}$$

may be described by the standard semiclassical Maxwell-Bloch equations [1, 4-7, 22] for the dimensionless complex amplitudes of the counter-propagating waves, $a_\pm = \tilde{a}_\pm \bar{\varepsilon} / (2\pi d N_0)$, complex spectral density of the polarization of an active medium, $p_\pm = P_\pm / (d N_0 f(\Delta))$, and related two components of the inversion of energy levels, namely, a slowly varying in space (real) component, $n(\Delta)$, and a half-wavelength modulated (complex) one, $n_z(\Delta)$, originating from the self-consistent beating of the counter-propagating waves. A similar half-wavelength modulated component of the polarization is ignored for simplicity's sake as it plays a minor role under the condition of the strong inhomogeneous broadening of the laser transition. Using slowly varying dimensionless amplitudes, $a_\pm(\varsigma,\tau)$, and taking into account the Bragg coupling of counter-propagating waves due to the spatial modulation of a host dielectric permittivity, $\varepsilon = \bar{\varepsilon}\,\text{Re}[1 + 4\beta\sqrt{I}\exp(2i\varsigma)]$, one can write down the Maxwell equations in a following form

$$\left[\frac{\partial}{\partial \tau} \pm \frac{\partial}{\partial \varsigma}\right] a_\pm = i\beta a_\mp + \frac{i}{\sqrt{I}} \int_{-\infty}^{\infty} p_\pm(\Delta) f(\Delta) d\Delta, \tag{2}$$

where $\Delta = (\omega - \omega_0)/v_c$ is a normalized frequency shift from a frequency of the Bragg resonance ω_0; a parameter $I = v_c^2 / \omega_{21}^2$ is small, $I \ll 1$; $\omega_{21} \approx \omega_0$ is a central frequency of the spectral line of the active medium, which is assumed to have the inhomogeneous broadening of a Lorentz type (symmetric with respect to the Bragg resonance), $f(\Delta) = \Delta_0 / \pi(\Delta^2 + \Delta_0^2)$; $\tau = t v_c$ and $\varsigma = z v_c \sqrt{\bar{\varepsilon}} / c$ are dimensionless time and coordinate; $L = B / B_c$ is a cavity length, B, normalized by means of a cooperative length, $B_c = c/(v_c \sqrt{\bar{\varepsilon}})$; $\Gamma_{1,2} = 1/(v_c T_{1,2})$ are dimensionless relaxation rates of the inversion and polarization of an active center. The boundary conditions describe reflections of the field at the laser facets with a given complex reflection factor, $\sqrt{R}$ (related to an amplitude a).

3. Traditional empirical modes

The simplest generalization of the known approach of the cold or hot modes [21, 22] is a decomposition of the fields $a_\pm(\varsigma,\tau)$, taken for a dense set of equally spaced grid points along a cavity ($\varsigma_j / \Delta\varsigma = 0,1,2,...,D$, where $\Delta\varsigma = L/D$ is an array pitch which is equal to 0.1 in the following calculations), by means of the time independent *Complex Empirical Orthogonal Functions* (CEOFs), $v_{\pm i}(\varsigma_j)$, which are defined as the normalized eigenvectors of a time-averaged $(D+1)\times(D+1) -$ covariance matrix, $\left\langle a_\pm(\varsigma_j,\tau) a_\pm^*(\varsigma_l,\tau)\right\rangle_\tau$, composed of the gridded data set (the indices j and l

enumerate the rows and columns, respectively) [16, 19, 20]: $a_{\pm}\left(\varsigma_j,\tau\right)=\sum_{i=1}^{D+1}Y_{\pm}^{(i)}\left(\tau\right)v_{\pm i}\left(\varsigma_j\right)$. An averaging is taken over the whole duration of the lasing under consideration (in what follows, it is equal to 16000 units of the dimensionless time τ), and the temporal dynamics is represented by the coefficients $Y_{\pm}^{(i)}$ in that decomposition.

Although the described linear decomposition procedure takes into account only the snap-short correlations of the field at any pair of grid points, it may be sufficient for the analysis of some phenomena in the lasers with low-Q cavities. For instance, if the reflections from the laser facets are not important ($|R|\ll 1$), the Bragg (DFB) reflections are moderate (so that their characteristic parameter is bounded as $b=\beta L\le\pi$), and the pumping is not too high, the superradiant lasing will contain few hot modes and demonstrate bunches of pulses with short timescales less or of the order of T_E and T_2, but may be quasi-periodic with a long bunch-repetition interval of the order of the time if an inversion repopulation by pumping, T_1, and, in a whole, will be described by even less number (maybe, one) of CEOFs with quasi-periodic amplitudes $Y_{\pm}^{(i)}(\tau)$.

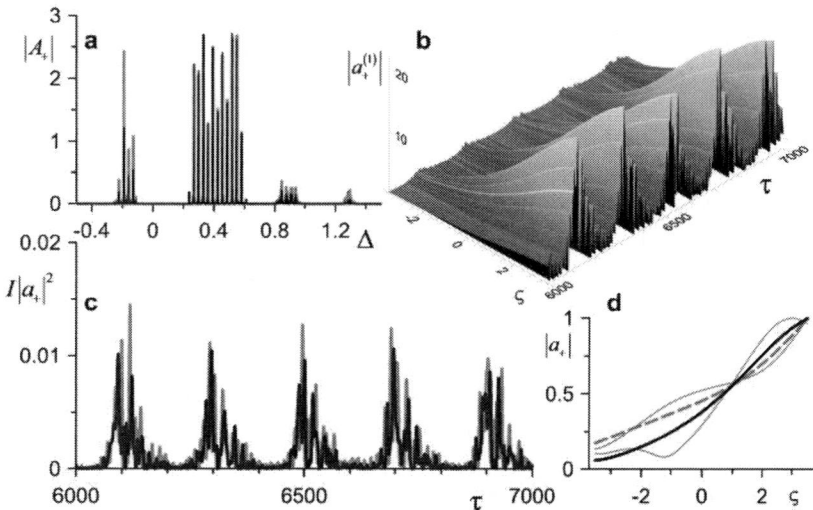

Figure 1. An example of CEOF which describes well the laser emission and fails to depict the field inside a cavity. (a) The spectra of the fields, $|A_+|$, and (c) the corresponding oscillograms of the intensities, $I|a_+|^2$, for a right-propagating wave of the total field (the blue lines) and the field of the main CEOF (the black lines) at a laser facet. (b) A space-time dynamics of the amplitude of this wave $|a_+^{(1)}|$ in the main CEOF. (d) The profiles of the right-propagating wave of the total field (two thin blue lines), the main CEOF (a black line), and the most unstable hot mode (a dashed orange line), which is calculated for the fixed maximum inversion, $n(\Delta)=1$. All profiles are normalized to their maximum values. The laser parameters are: $L=7$, $\sqrt{R}=0.1$, $\Delta_0=4$, $b=0.7$, $\Gamma_1=0.01$, $\Gamma_2=0.03$, $I=25\cdot 10^{-6}$.

An example is given in figure 1, where an emission of a laser with the relaxation parameters $T_1\approx 3T_2\approx 20T_E$ is well described by a single (main) CEOF, $a_{\pm}^{(1)}=Y_{\pm}^{(1)}v_{\pm 1}$ (see a plot 1b and the black lines in the plots 1a, c). In fact, it plays a part of a new hot mode, which can be defined with the use of a time-averaged spatial profile of population inversion. A field profile of this CEOF differs essentially from that of the main unstable hot mode (which is found for a homogeneous inversion of the active medium and shown by the dashed red line in a plot 1d) and makes it possible to characterize an output radiation with a 90% accuracy in energy without use of any STEMs. The description of an outgoing emission may be satisfactory for a large domain of parameters not too far from a lasing threshold. Nevertheless, the field patterns inside a cavity are varied as compared to the main CEOF pattern and need to be analyzed on the basis of the more adequate STEMs (cf. the thin blue lines, which show the field profiles at two moments of time when the maximum superradiant emission is achieved, and the thick black line, which shows the main CEOF, in the plot 1d).

Moreover, if the pumping is rather strong and/or the facet reflections $\sqrt{R}$ are not too weak (as compared to 1 and/or $\tanh(b)$), the several more or less independent sequences of superradiant pulses with different pulse-repetition intervals may be formed due to the oscillations of neighbouring hot modes. Also, a regime of a partial self-locking of the quasi-stationary generating modes with quite uniform frequency spacing at the periphery of the lasing spectrum is possible. It would result in a formation of one more quasi-periodic sequence of pulses with a pulse-repetition interval which is approximately equal or two times less than a cavity round-trip time, i.e., of the order of the value T_E. Bearing all these in mind, one can expect that an essential part of CEOFs will differ from the original cold modes of a cavity or hot modes of a laser, and the whole CEOF approach will be inefficient for the theoretical investigation of the superradiant-laser dynamics.

4. Time-dependent empirical modes

Thus, in many cases a proper description of both the field inside a cavity and the output emission requires a further generalization of the empirical-orthogonal-function approach beyond the CEOF technique. Namely, it is required not only spatial, but also temporal correlation analysis of the gridded data set of complex fields $a_\pm(\varsigma,\tau)$, taken at the discrete points along a cavity ($\varsigma_j / \Delta\varsigma = 0,1,2,...,D$, where $\Delta\varsigma = L/D$ is an array pitch) at the discrete moments of time ($\tau_k / \Delta\tau = 0,1,2,...,\mathrm{M}$, where $\Delta\tau = T/M$ is a time step equal to 0.1 in our simulations) within a common time interval $[\tau, \tau+T]$, which is defined by a time scale T under investigation. Systematic investigation of the multi-scale temporal dynamics may even require a use of several series of different generalized empirical modes with different time scales (relevant to various processes), e.g., the averaged periods of generation of the superradiant pulses of various types. In general, a proposed approach follows the techniques of a multidimensional spectral analysis [19, 25, 26] and a compact representation of spatial-temporal data sets [20]. Our approach to the study of the steady or slow varying regimes of lasing originates from the field decompositions (related to different above-mentioned time scales) over the so-called *Space-Time Complex Empirical Orthogonal Functions* (STCEOFs). According to a definition given below, STCEOFs are normalized eigenvectors of the above-mentioned time-averaged extended covariance matrix of a given gridded set of the time-shifted field data capable of taking into account the delayed interparticle interaction, which is owing to the physical processes in the active medium and plays a leading part in many cases. In what follows, depending on a lasing problem and generalizing the standard cold or hot modes, we will introduce also *the Space-Time Empirical Modes (STEMs)* as some individual STCEOFs (including their amplitudes as factors) or some particular steady-state superpositions of several STCEOFs (again, including their amplitudes as factors).

In the following examples, for definiteness, such an approach is performed for a single time scale, $T \approx L$, chosen close to a half cavity round-trip time. The STCEOF construction is carried out separately for each counter-propagating field, $a_\pm(\varsigma,\tau)$, represented by the above-mentioned gridded time series. Specifically, for an arbitrary time, τ, we write down a complex column vector, $(a_\alpha(\tau))$, which consists of $(D+1)(M+1)$ elements (enumerated by an index α) grouped as a sequence of the consecutive $M+1$ snap-shorts (maps) of the field (at the moments τ_k), $a(\varsigma_j, \tau_k)$, each given by a $(D+1)-$ column vector (an index j enumerates mesh nodes along a cavity).

The index α may be used in a dual form, $j \mapsto k$, which clearly shows that the consecutive snap-shorts (i.e., $D+1$ values of the field in the grid points $j=0, j=1,\ldots,j=D$) are arranged in a time series according to the growing counts via index $k=0,1,\ldots,\mathrm{M}$: $(a_\alpha(\tau)) \equiv (a_{j \mapsto k}(\tau))$, i.e., $(a_\alpha)^T \equiv (a_{0 \mapsto 0}, a_{1 \mapsto 0}, \ldots, a_{D \mapsto 0}, a_{0 \mapsto 1}, a_{1 \mapsto 1}, \ldots, a_{D \mapsto 1}, \ldots, a_{0 \mapsto M}, a_{1 \mapsto M}, \ldots, a_{D \mapsto M})^T$. What we have to find as the non-stationary STCEOFs are the normalized $(D+1)(M+1)$ – eigenvectors, v_i, of an extended Hermitian covariance matrix, $\langle a_\alpha(\tau) a_\beta^*(\tau) \rangle_\tau$ (α and β enumerate the rows and columns, respectively), which results from the time-averaging of the above-mentioned gridded time series: $(v_i)^T = \left(\left(v_i^{j \mapsto 0}\right)^T, \left(v_i^{j \mapsto 1}\right)^T, \ldots, \left(v_i^{j \mapsto M}\right)^T \right)$, where a $(D+1)$ – column vector, $v_i^{j \mapsto k}$, is introduced for the given indices k and i. These eigenvectors v_i contain an important information on a 'fast' dynamics of fields, including the time scales up to the given time scale, T.

Complete dynamics of the fields, including also a 'slow' one with the time scales greater than T, may be seen from a full decomposition of each counter-propagating wave by means of the STCEOF basis and two types of the following expansion functions of time, i.e., the time-dependent complex amplitudes, $Y^{(i)}(\tau)$ and $Z^{(i)}(\tau) = \left(a(\tau) v_i^*\right) \equiv \sum_{\alpha=1}^{(D+1)(M+1)} a_\alpha(\tau) \cdot v_i^{*\alpha}$ (for details of a decomposition algorithm, see [25]):

$$a(\varsigma_j, \tau) \equiv \sum_{i=1}^{(D+1)(M+1)} Y^{(i)}(\tau) a^{(i)}(\varsigma_j, \tau) = \sum_{i=1}^{(D+1)(M+1)} \frac{1}{M+1} \sum_{k=0}^{M} Z^{(i)}(\tau - \tau_k) v_i^{j \mapsto k}, \tag{3}$$

where $\tau \geq T$. Here $Y^{(i)}(\tau)$ are the complex amplitudes of the normalized spatial profiles of STCEOFs, $a^{(i)}(\varsigma_j, \tau)$, which are defined by the above time-averaging of a projection of the gridded time series onto the eigenvectors v_i within the time interval T. The terms in the above decomposition of field over the STCEOF basis, $a^{(i)}(\varsigma_j, \tau)$, like in the similar decomposition over the CEOF basis, $v_{\pm i}(\varsigma_j)$, are arranged according to the decreasing order of eigenvalues corresponding to the eigenvectors v_i, since each eigenvalue of an extended covariance matrix is proportional to a relative power of a given STCEOF with respect to a power of the complete laser field. The time scales of these STCEOFs (including their complex amplitudes), $Y^{(i)}(\tau) a^{(i)}(\varsigma_j, \tau)$, are well defined, different and, as a rule, become shorter and shorter for higher numbers i, so that the first STCEOFs are responsible for slower dynamics than subsequent ones. We expect that, to pick out qualitatively the main features of the field dynamics in a superradiant laser, it is sufficient to consider several first STCEOFs, which contain, say, 90% or 99 % fraction of the power of laser field.

A field at a laser facet, for instance, at the point $\varsigma = 0$, is equal to the same superposition of all STCEOFs (if the reflection factor, $\sqrt{R}$, is taking into account), so that each STCEOF contribution is defined by a product of the time-dependent complex amplitude, $Y^{(i)}(\tau)$, and the known function of time, $a_-^{(i)}(\varsigma = 0, \tau)$. For an opposite laser facet, a field of the counter-propagating wave at the point $\varsigma = L$ consists of the sum, $\sum_{i=1}^{(D+1)(M+1)} Y_+^{(i)}(\tau) a_+^{(i)}(\varsigma = L, \tau)$, of all similar STCEOFs, $a_+^{(i)}(\varsigma = L, \tau)$.

5. Examples of the space-time empirical modes

Figures 2 and 3 show an example of a quasi-chaotic emission of a sequence of superradiant pulses originated from a cooperative dynamics of the several different spectral subensembles of active

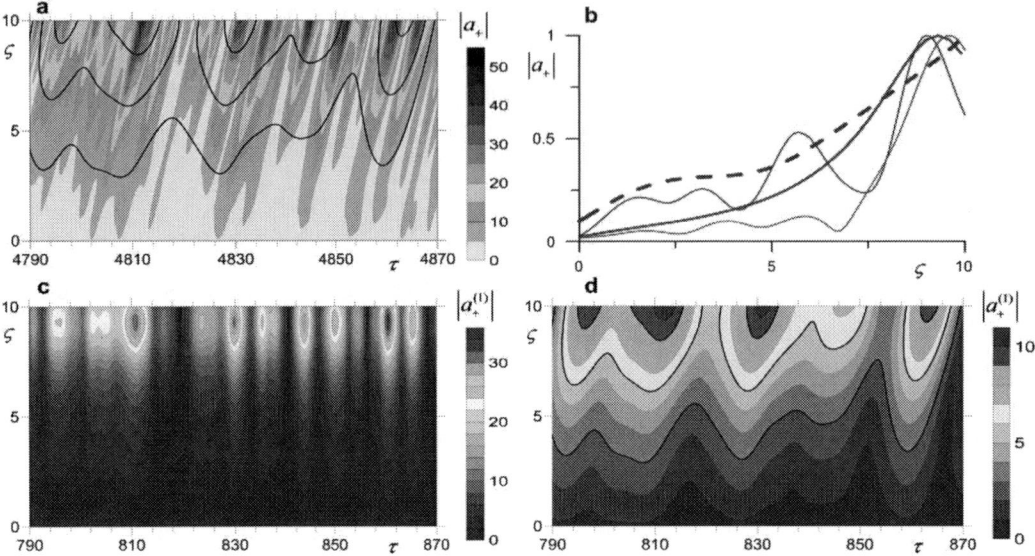

Figure 2. A spatial-temporal structure of the contribution of the right-propagating wave, $|a_+|$, to (a) the total field, (c) the field of the main CEOF, and (d) the field of the first STCEOF defined with a use of a time scale $T = 12.5$. (b) The typical profiles of the right-propagating wave within the total field (two thin colour lines), the main CEOF (a blue line), and the most unstable hot mode (a dashed black line) calculated for the fixed maximum inversion, $n(\Delta) = 1$. All profiles are normalized to their maximum values. The laser parameters are: $L = 10$, $\sqrt{R} = 0.1e^{i\pi/2}$, $\Delta_0 = 4$, $b = 1$, $\Gamma_1 = 0.01$, $\Gamma_2 = 0.02$, $I = 25 \cdot 10^{-6}$. The contours of a plot (d) are put onto a plot (a) in order to compare the dynamics of the total field ($|a_+|$) and the field of the first STCEOF ($|a_+^{(1)}|$).

centers, or, in other words, caused by the instabilities of several different hot modes of the laser. According to figure 2, a travelling-wave nature of the total field is well described by the first STCEOF which contributes about a quarter of the total power. A field of the main CEOF, which contributes a bit higher than a half of the total power, has nothing to do with this travelling-wave phenomenon and, in fact, contains an information on some averaged field profile in the cavity only. It is useful to combine two first STCEOFs in one STEM. Then, the latter contains a bit less than a half of the total field power, covers a well-defined range of spectrum (see figure 3b), and yields a precise, complete description of the sequence of the superradiant pulses emitted by a subensemble of the active centers which have spectral localization in a vicinity of the photonic bandgap (owing to the Bragg DFB) and are closely related to the most unstable two hot modes situated at the bandgap edges. A typical duration of these superradiant pulses is of the order of the cavity lifetime, $T_E \approx 8$, and much less than the relaxation time of polarization, $T_2 = 50$. Similarly, the next STCEOFs make it possible to describe other sequences of superradiant pulses emitted by the neighbouring spectral subensembles of active centers.

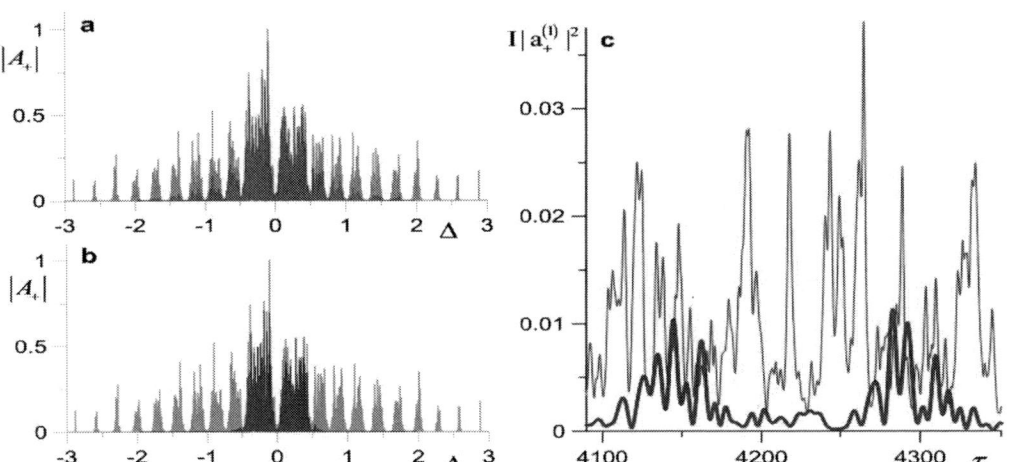

Figure 3. The spectra of the total field (the grey lines on the plots (a) and (b)), the field of the main CEOF (a red line on a plot (a)), and the main STEM formed by a superposition of the first two STCEOFs (a green line on a plot (b)). A comparison of the oscillograms of an intensity, $I \mid a_+^{(1)} \mid^2$, of the field of the main CEOF (a red line on a plot (c)) and the main STEM (a green line on a plot (c)) is also shown. The laser parameters are: $L = 10$, $\sqrt{R} = 0.1 e^{i\pi/2}$, $\Delta_0 = 4$, $b = 1$, $\Gamma_1 = 0.01$, $\Gamma_2 = 0.02$, $I = 25 \cdot 10^{-6}$.

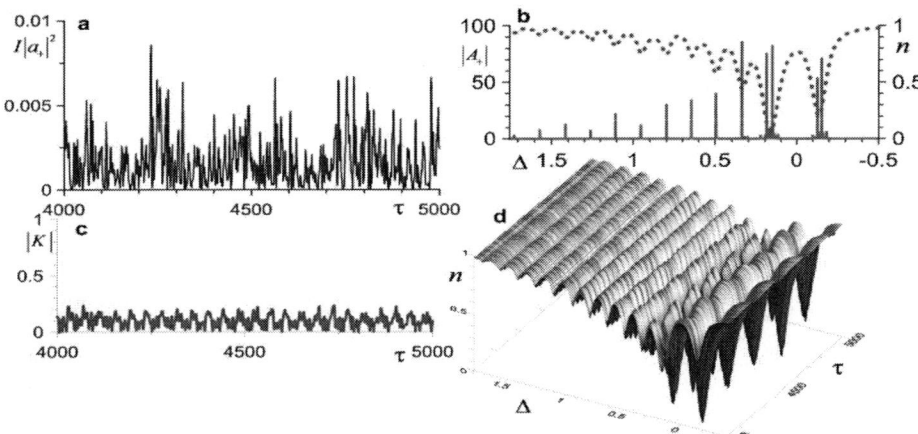

Figure 4. (a) An oscillogram of the intensity, $I \mid a_+ \mid^2$, of a right-propagating wave. (b) A spectral distribution of the population inversion, $n(\Delta)$, (at a moment of time when a superradiant pulse is emitted) and a spectrum of the field, $\mid A_+ \mid$, at a laser facet. (c) A correlation function, $\mid K \mid = \left| \int_{-\bar{T}/2}^{\bar{T}/2} a(t) a^*(t+\tau) dt \right| \Big/ \int_{-\bar{T}/2}^{\bar{T}/2} \mid a(t) \mid^2 dt$, of the field at a laser facet for a time-averaging interval much greater than a typical interval between the bunches of superradiant pulses ($\bar{T} \approx 1000$). (d) A dynamical spectrum of the population inversion of the lasing transition. The laser parameters are: $L = 20$, $\sqrt{R} = 0.1$, $\Delta_0 = 13$, $b = \sqrt{3}$, $\Gamma_1 = 0.01$, $\Gamma_2 = 0.03$, $I = 2.3 \cdot 10^{-6}$.

A STEM-based analysis provides a deep qualitative insight into the straightforward numerical solution to the integral-differential equations of a superradiant-laser dynamics [21-24, 27-30] and makes it possible to pick out the important features of the above-mentioned regimes of oscillations under CW pumping. In a rest part of the paper, we will consider one such regime which takes place

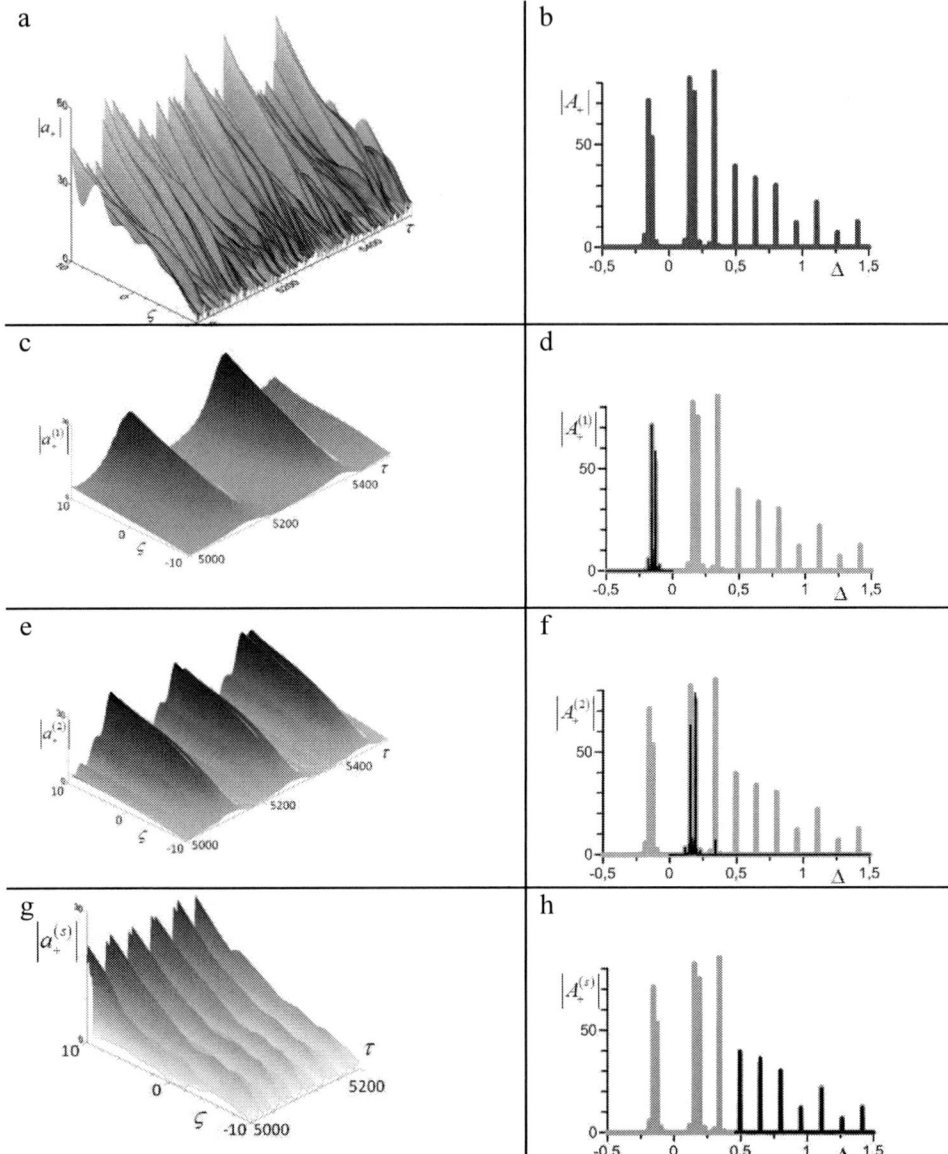

Figure 5. The spatial-temporal dynamics of the amplitude, $|a_+|$, of the right-propagating field and the spectrum of this field, $|A_+|$, at a laser facet in the case of the multimode oscillations under CW pumping. (a), (b) The total field and its spectrum. (c), (d) The field and spectrum of the first STCEOF which describes one independent superradiant mode (shown by black on a grey background of the total spectrum). (e), (f) The field and spectrum of the second STCEOF which describes another superradiant mode and demonstrates a nonlinear interaction with the quasi-stationary lasing modes from a right-hand side of the total spectrum. (g), (h) The field and spectrum of a combined STEM which is defined as a superposition of seven STCEOFs, from 4th to 10th ones (again shown by black), and represents the main part of a pulse formed due to the self-mode-locking effect. All STCEOFs are calculated with a use of a time scale $T = 20$. The laser parameters are: $L = 20$, $\sqrt{R} = 0.1$, $\Delta_0 = 13$, $b = \sqrt{3}$, $\Gamma_1 = 0.01$, $\Gamma_2 = 0.03$, $I = 2.3 \cdot 10^{-6}$.

not far from a superradiant threshold in the case when a spectrum of the excited hot modes of a low-Q combined DFB – Fabry-Perot cavity is enriched due to a mutual action of the weak reflections at the laser facets and a nonlinear coupling between the Fourier components of the superradiant field outside the photonic band gap and the spectrally remote quasi-stationary laser modes. Then, in general, the laser radiation has a form of a nonlinear superposition of a quasi-periodic or quasi-chaotic sequence of the ultrashort powerful superradiant pulses (figure 4a) and a quasi-periodic (more regular) sequence of the comparable pulses formed by the quasi-stationary self-locked modes of the Fabry-Perot cavity with an equidistant spectrum (figure 4b). The latter are locked, that is obey certain phase relations, due to an effect of saturation of a pulse absorption which takes place in the presence of the deep spectral holes of population inversion (figure 4b, a dashed line, and figure 4d) as it may tend to zero or even become negative during the action of the superradiant pulses.

In the above example, a degree of correlation of the quasi-chaotic radiation at a laser facet is not high (less than 25% in figure 4c) because there are continuous beatings between some pulses of more or less independent origin which are comparable in power and have different time scales. Mainly, these are the partially formed superradiant pulses which are generated in the two spectral channels at the edges of the photonic band gap and have a duration of the order of the relaxation time of polarization, $T_2 \approx 1.5 T_E \approx 34$. A repetition interval is on the order of time it takes the pumping to restore a high level of the population inversion, $T_1 = 100$. These pulses interfere also with the self-mode-locked pulse which is travelling (and distributed) around a cavity and responsible for a formation of a sequence of more frequent emission pulses with an average repetition interval $T \approx L = 20$ and even shorter duration less than the above-mentioned relaxation times.

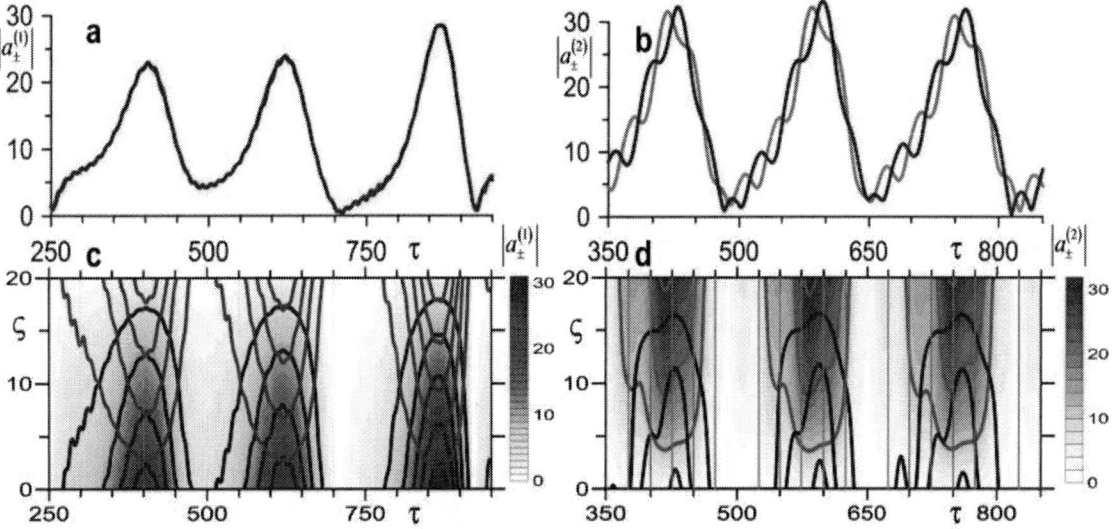

Figure 6. (a), (b) The oscillograms of the amplitudes of the counter-propagating waves, $|a_\pm|$, (at the opposite laser facets) of the first and second superradiant STEMs defined as the first and second STCEOFs, respectively. (c), (d) The spatial-temporal dynamics of the amplitudes of these waves, $|a_\pm(\varsigma, \tau)|$. The laser parameters are: $L = 20$, $\sqrt{R} = 0.1$, $\Delta_0 = 13$, $b = \sqrt{3}$, $\Gamma_1 = 0.01$, $\Gamma_2 = 0.03$, $I = 2.3 \cdot 10^{-6}$.

The unique simultaneous generation of the two pulse sequences comparable in power and different in the spectral and temporal scales (both may differ by many times) is worth to analyze in detail by means of the universal STCEOF basis (figure 5). Note that a straightforward modeling of the

Maxwell-Bloch equations does not allow one to separate the space-time structures of individual modes under consideration, and an actual experiment has nothing to say on this account as well.

According to figure 5, the discussed complicated quasi-chaotic oscillations (shown in figures 4a and 5a), in fact, result from a combination of three quite regular dynamics separated in spectrum and described by a pair of the superradiant STEMs, which correspond to the 1st and 2nd STCEOFs, respectively, (figure 5c, e, figure 6) and an additional combined STEM, which is formed by a group of the 4th to 10th STCEOFs (figure 5g, figure 7) and represents the main part of the self-mode-locked pulse. An interaction of the latter with the two superradiant modes is mediated by third STCEOF which is not shown in the left column of figure 5.

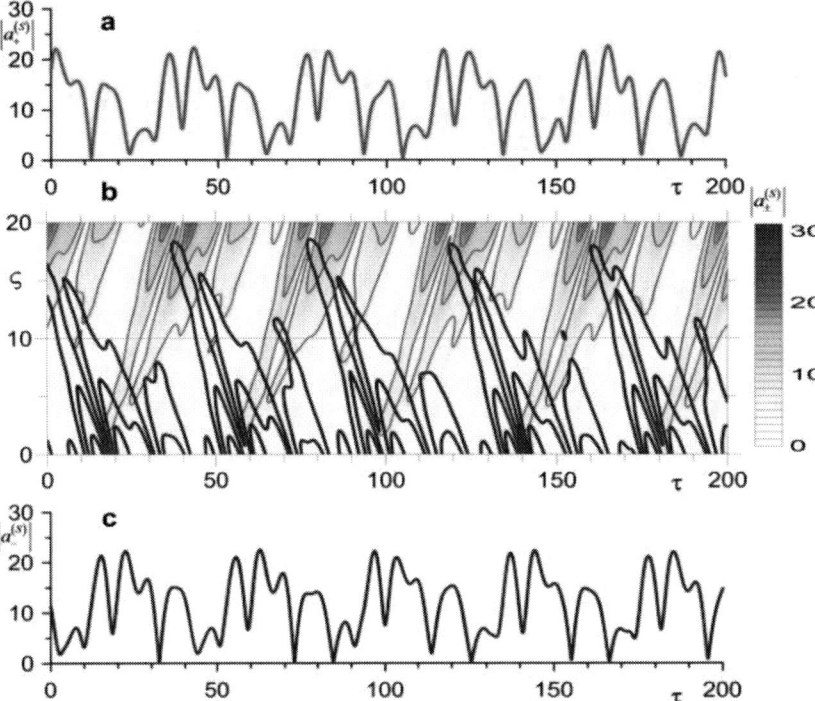

Figure 7. The oscillograms of the amplitudes of the counter-propagating waves, $|a_\pm^{(s)}|$, of the combined STEM (see a right-hand side of the spectrum in Fig. 5h) related to the main part of the round-travelling pulse which is formed by the self-locked hot modes and produces an output radiation from the right (a, a red line) and left (c, a black line) laser facets. (b) The spatial-temporal dynamics of the amplitudes of the right-propagating (the red contours) and left-propagating (the black contours) waves, $|a_\pm^{(s)}(\varsigma,\tau)|$, of the same STEM. The laser parameters are: $L = 20$, $\sqrt{R} = 0.1$, $\Delta_0 = 13$, $b = \sqrt{3}$, $\Gamma_1 = 0.01$, $\Gamma_2 = 0.03$, $I = 2.3 \cdot 10^{-6}$.

The superradiant STEMs, each of them consists of two counter-propagating waves with a smooth (within a cavity length scale) non-stationary spatial structure, are emitted simultaneously from the opposite laser facets (see two pairs of curves, $a^{(1)}(\tau)$ and $a^{(2)}(\tau)$, in figure 6a, b and, hence, have a standing-wave (mode) character (a shallow ripple in figure 6b and a weak phasing-out of the amplitude oscillations in figure 6d are not important statistically). The time scales of the first two

STEMs differ essentially ($\tau_1 = 250$ and $\tau_2 = 150$, respectively). The third (combined) STEM, $a^{(S)}(\tau)$, consists of the fields of seven STCEOFs, each of them contributes to a spectrum of two or three different hot modes of the laser. This STEM describes a spatial-temporal structure of the self-mode-locked pulses which are periodic in time and extend over the whole cavity in such a way that they repeat themselves after every successive reflection at each laser facet. The counter-propagating waves, present in these pulses and shown in figure 7b by the red and black contours, re-emit each other continuously when moving across the Bragg periodic lattice of the dielectric permittivity in the DFB – Fabry-Perot cavity. It is the unique combine STEM that clarifies the complicated spatial-temporal structure of the round-travelling self-mode-locked pulses.

6. Conclusions

The presented examples of the space-time empirical modes (STEMs) prove that these modes are useful for the analysis of complicated features of a strongly non-stationary field which are typical for the superradiant lasing in the low-Q cavity. Thus, the suggested STEM approach is an efficient tool in the dynamical theory and interpretation of various regimes of the superradiant lasers as well as in the development of their applications in the optical information processing, the wideband dynamical spectroscopy, and the diagnostics of the many-particle processes in condensed active media.

Acknowledgements

The work was partially supported by the Russian Foundation for Basic Research (a grant # 16-02-00714) and the Program of Fundamental Research of the Physical Science Brunch of the Russian Academy of Sciences (a project # III.7 "Fundamentals and experimental development of the perspective semiconductor lasers for the industrial and technological purposes") and the Competitiveness Program of National Research Nuclear University "MEPhI". One of the authors (V.V. Kocharovsky) is very grateful for the nice hospitality of MEPhI during his visit supported by the mentioned Program.

References

[1]   Khanin Ya. L. 2006 *Fundamentals of Laser Dynamics* (Cambridge: Cambridge International Science Publishing)

[2]   Arecchi F T and Harrison R G 2011 *Instabilities and Chaos in Quantum Optics* (London: Springer Verlag)

[3]   Ohtsubo J 2013 *Semiconductor Lasers: Stability, Instability and Chaos* (Series: Springer Series in Optical Sciences vol. 111)

[4]   Roldan E et al 2005 Trends in Spatiotemporal Dynamics in Laser. Instabilities, Polarization Dynamics, and Spatial Structures (Trivandrum: Research Signpost) (Preprint http://www.arXiv: physics/0412071V1)

[5]   Belyanin A, Kocharovsky V, Kocharovsky Vl 1997 Collective QED processes of electron-hole recombination and electron-positron annihilation in a strong magnetic field *Quantum and Semiclassical Optics* **9** 1

[6]   Zheleznyakov V, Kocharovskii V, Kocharovskii Vl 1989 Polarization waves and superradiance in active media *Sov. Phys. Usp.* **32** 835

[7]   Golubyatnikova E, Kocharovsky V, Kocharovsky Vl 1997 Mode Instability and Nonlinear Superradiance Phenomena in Open Fabry-Perot Cavity *International Journal of Computers and Mathematics with Applications* **34** 773

[8]   Qiao P et al 2013 Theory and experiment of submonolayer QD metal-cavity surface-emitting microlasers *Optics Express* **21** 30336

[9]   Ding C et al 2012 Observation of In-related collective spontaneous emission (superfluorescence) in Cd0.8Zn0.2Te:In crystal *Appl. Phys. Lett.* **101** 091115

[10]   Dai D and Monkman A 2011 Observation of superfluorescence from a quantum ensemble of coherent excitons in a ZnTe crystal: Evidence for spontaneous Bose-Einstein condensation

of excitons *Phys. Rev. B* **84** 115206

[11] Germann T et al 2008 High-power semiconductor disk laser based on InAs/GaAs submonolayer quantum dots *Appl. Phys. Lett.* **92** 101123

[12] Scheibner M et al 2007 Superradiance of quantum dots *Nature Physics* **3** 106

[13] Kim J et al 2013 Fermi-edge superfluorescence from a quantum-degenerate electron-hole gas *Scientific Rep.* **3** 3283

[14] Jho Y et al 2010 Cooperative recombination of electron-hole pairs in semiconductor quantum wells under quantizing magnetic fields *Phys. Rev. B* **81** 155314

[15] Kalinin P et al 2012 On the problem of lasing in traps for the Bose condensation of dipolar excitons *Semiconductors* **46** 1351

[16] Jollife I 1986 Principal Component Analysis (Springer)

[17] Hannachi A, Jolliffe I and Stephenson D 2007 Empirical orthogonal functions and related techniques in atmospheric science: A review *Int. J. Climatol.* **27** 1119

[18] Monahan A et al 2009 Empirical Orthogonal Functions: The Medium is the Message *Journal of Climate* **22** 6501

[19] Navarra A and Simoncini V 2010 *A Guide to Empirical Orthogonal Functions for Climate Data Analysis* (Springer Science+Business Media B.V.)

[20] Mukhin D et al 2015 Predicting critical transitions in ENSO models II: Spatially dependent models *Journal of Climate* **28** 1962

[21] Kocharovsky Vl et al 2013 Dynamics of the class D lasers based on the Bose-Einstein condensates, the submonolayer quantum dots and other exotic active media *Nonlinear waves'2012* (Nizhny Novgorod: IAP RAS) pp 398-428 (in Russian)

[22] Kocharovsky Vl et al 2015 Superradiant Lasing and Collective Dynamics of Active Centers with Polarization Lifetime Exceeding Photon Lifetime *Advanced Lasers: Laser Physics and Technology for Applied and Fundamental Science* ed. O Shulika (Series: Springer Series in Optical Sciences. V. 193) chapter 4 pp 49-69.

[23] Kocharovsky Vl et al 2010 Perspectives of creation of a superradiant heterolaser *Proc. II Symp. on coherent optical radiation of semiconductor materials and structure* (Moskow: Lebedev Institute of Physics, RAS) p 68 (*in Russian*)

[24] Kalinin P et al 2012 Features and coherence of pulses generated by the superradiant lasers based on the multilayered Bragg heterostructure with submonolayer quantum dots under the conditions of self-mode-locking of longitudinal modes *Proc. III Symp. on coherent optical radiation of semiconductor materials and structure* (Moskow: Lebedev Institute of Physics, RAS) p 71 (*in Russian*)

[25] Ghil M et al 2002 Advanced spectral methods for climatic time series *Rev. Geophys.* **40** 1

[26] Plaut G and Vautard R 1994 Spells of oscillations and weather regimes in the low-frequency dynamics of the Northern Hemisphere *J. Atmos. Sci.* **51** 210

[27] Kocharovskaya E et al 2015 The Dynamical Spectra of the Superradiant Heterolasers: Spatial-Temporal-Dependent Mode Technique Versus Cold or Hot Mode Techniques *Tech. Digest Int. Workshop "Nonlinear Photonics: Theory, Materials, Applications" NPh-15* pp 46-7

[28] Loskutov E et al 2015 Spatial-temporal empirical modes as an instrument of studying superradiant laser dynamics *Thesis Int. Workshop "DyNeMo-Clim"* p 34

[29] Kocharovsky Vl and Kocharovsky V 2016 Progress and perspectives of the superradiant lasers *Proc. XX Int. Symposium "Nanophysics and Nanoelectronics"* vol 2 p 632 (*in Russian*)

[30] Kocharovskaya E et al 2015 Superradiant semiconductor distributed-feedback lasers. An analysis of dependence of a laser dynamics on a relation of between the relaxation times of a field and a polarization of active medium *Proc. XX Int. Symposium "Nanophysics and Nanoelectronics"* vol 2 p 630 (*in Russian*)

InAsP/AlGaInP/GaAs QD laser operating at ~770 nm

A B Krysa[1,2], J S Roberts[1], J Devenson[1], R Beanland[3], I Karomi[4,5], S Shutts[4] and P M Smowton[4]

[1] EPSRC National Centre for III-V Technologies, Department of Electronic and Electrical Engineering, University of Sheffield, Mappin Street, Sheffield, S1 3JD, UK
[2] P.N. Lebedev Physical Institute, Russian Academy of Sciences, 53 Leninskiy pr., Moscow, 119991, Russia
[3] Department of Physics and Astronomy, University of Warwick, Coventry, CV4 7AL, UK
[4] Physics and Astronomy, Queens Building, The Parade, Cardiff, CF24 3AA, UK
[5] University of Mosul, Mosul, Iraq

E-mail: a.krysa@sheffield.ac.uk

Abstract. We present a study of metalorganic vapour phase epitaxy of ternary InAsP quantum dots in AlGaInP/GaAs for application in laser diodes. The properties of InAsP QD laser structures were compared with reference samples containing binary InP QDs. Based on X-ray diffraction, the molar fraction of arsenic in InAsP QDs was estimated to be ~25%. Room temperature liquid contact electro-luminescence measurements revealed a long wavelength shift of the InAsP QD emission to ~775 nm as compared with the InP QD emission at 716 nm and an increased full width at half maximum of the spontaneous emission (71 meV vs 50 meV). As cleaved, 4 mm long and 50 μm wide InAsP QD lasers operated in a pulsed regime at room temperature at ~770 nm with a threshold current density of 155 A/cm^2 and a maximum output optical power of at least 200 mW. The maximum operation temperature was at least 380 K.

1. Introduction

Since the publications of Ahopelto et al [1], DenBaars *et al* [2] and Carlsson *et al* [3], self-assembled InP quantum dots (QDs) grown in (Al)GaInP matrices on GaAs substrates have attracted much interest as an alternative materials system for optoelectronic applications, primarily laser diodes [4, 5, 6], in the red – near infra-red spectral band. Recently, a significant improvement of the threshold current in InP QD laser diodes has been achieved [7, 8, 9], and other devices incorporating InP QDs such as, semiconductor disc lasers [10, 12], dual-wavelength laser diodes [12, 13], optical amplifiers with ultrafast gain dynamics [14[14]], saturable absorber mirrors for mode-locking [15, 16] and single-photon emitters [17, 18, 19, 20, 21] have been demonstrated. Also, rich nuclear spin phenomena have been observed in InP QDs [22, 23, 24, 25, 26], and efficient optical pumping of nuclear spin polarisation with ultra-long depolarisation times up to ~6000 s has been demonstrated [27]. Further to these studies, photonic crystal cavities comprising InP QDs have been fabricated [28], which would enable realisation of optically controllable spin qubits.

The operating wavelengths in the above examples can be controlled by the QD size, composition of the surrounding AlGaInP matrix and atomic diffusion between the dots and surrounding matrix, and has certain limits (here, we do not consider temperature driven spectral shift neither specific device

Content from this work may be used under the terms of the Creative Commons Attribution 3.0 licence. Any further distribution of this work must maintain attribution to the author(s) and the title of the work, journal citation and DOI.
Published under licence by IOP Publishing Ltd

fabrication techniques, e.g. DFB grating [13]). In particular, a longer operation wavelength can be achieved by growing larger QDs. However, formation of misfit dislocation and difficulties with planarization of the epitaxial surface after growth of large QDs [29] limit the operation wavelength of InP QD lasers to around 750 nm at room temperature (RT).

Adding arsenic to InP QDs lowers the bandgap of the dot material and increases the lattice mismatch in respect to GaAs substrates, and, thus, offers extra flexibility in engineering the above QD structures with a potential of extending their spectral operation range towards longer wavelengths in the near infra-red band.

Growth of InAsP QDs has been attempted earlier. Vinokurov *et al* [30] reported InAsP QD growth by MOVPE using a mixture of arsine and phosphine but did not observe the expected long wavelength spectral shift of the QD emission, as they presumed, due to formation of smaller size InAsP QDs in comparison to the growth of nominally binary InP QDs. On contrast, Fuchi *et al* [31] observed a longer wavelength emission from InAsP QDs when used a droplet hetero-epitaxy technique with tertiarybutyl precursors for the group V elements and a sophisticated reagents' switching procedure.

In this research, we applied an MOVPE process with the elements which are common for manufacturing red lasers [32, 33, 34], including commonly used group V precursors, i.e. arsine and phosphine, and trimethyl-metal precursors, for InAsP QD growth. We have demonstrated the feasibility of this approach to extend the operation of InP QD lasers towards longer wavelengths at least to ~770 nm.

2. Epitaxial growth and calibrations of P and As fractions in QDs

The epitaxy of laser structures was performed on (100) GaAs substrates with a miscut angle of 10° towards <111>A in a low pressure (150 Torr) horizontal flow reactor. The growth temperatures measured by a thermocouple inside the graphite susceptor were 710 $^{\circ}$C or 730 $^{\circ}$C. In addition to the groups V and III precursors mentioned in the introduction, we also used disilane and dimethylzinc for n- and p-type doping, respectively.

The active region of the laser structures consisted of 5 InP or InAsP QD sheets with a GaInP quantum well with a thickness of 8 nm grown above each QD sheet and separated by a layer of $(Al_{0.3}Ga_{0.7})_{0.52}In_{0.48}P$ with a thickness of 16 nm. The active region was sandwiched with an $Al_{0.52}In_{0.48}P$ clad and $(Al_{0.3}Ga_{0.7})_{0.52}In_{0.48}P$ core waveguide with a p-GaAs contact layer on the top. Throughout the growth of the active region, the flow of phosphine was kept constant at 300 sccm, and arsine (6.25 sccm) was introduced to the reactor during growth of InAsP QDs.

To evaluate the molar fraction of arsenic in the InAsP QD layers, a calibration sample comprising of 20 periods of alternating InAsP (3.2 nm) and InP (29.2 nm) layers was grown on on-axis (100) InP substrates and analysed by a Bede QC2a X-ray diffractometer around the (004) plane. The reason for using such InP-InAsP crystalline superlattice (SL) rather than bulk InAsP layers in our calibrations was to "dilute" the average strain in the grown structures and, consequently, to avoid the complications in X-ray diffraction analysis due to the strain relaxation in the crystalline lattice.

Similarly to the InAsP QD laser active region, the calibration sample was grown at 710 $^{\circ}$C and the same constant phosphine flow. Arsine was switched to the reactor during growth of InAsP layers. A theta-2theta diffraction scan of the calibration sample is presented in figure 1, bottom curve. The "zero" order peak of the SL is offset towards smaller diffraction angles by 216" which corresponds to a moderate compressive strain of 0.08%. Assuming no strain relaxation and abrupt reagents' switching, the calculated diffraction spectrum (bottom curve) resulted in a good agreement with the measured one at an arsenic fraction of ~25% in the group V crystalline sublattice. The corresponding decrease of the bandgap energy of InAsP is ~250 meV.

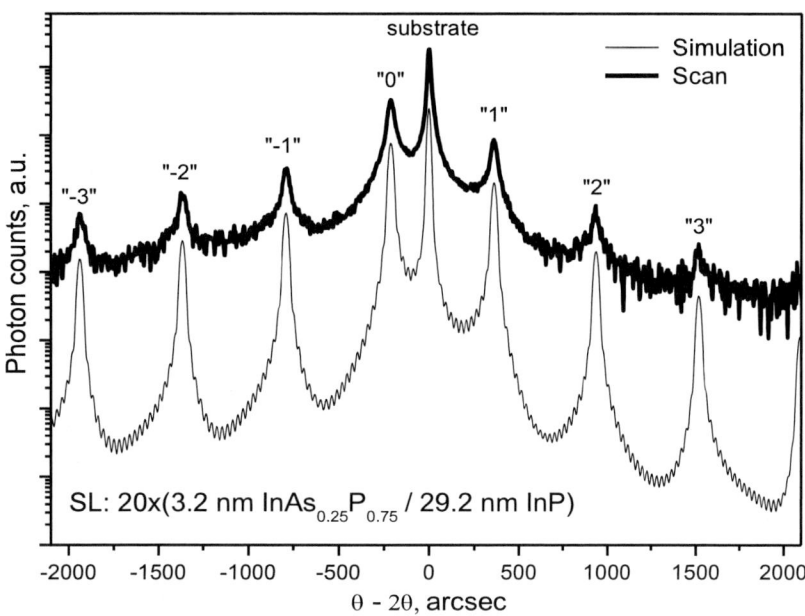

Figure 1. Theta-2theta scans of InP/InAsP SL: measured (top) and simulated (bottom) assuming an arsenic fraction of 25%.

3. Liquid contact electro-luminescence measurements

Prior to device processing, the as grown epi-wafers were examined using pulsed liquid contact electro-luminescence (LCEL). The principles of the LCEL technique for characterization of laser diodes were formulated and realized by Zory et al [35, 36, 37]. In our earlier work [38], we have successfully used LCEL for characterization of InP QD lasers. The setup used in the current research was built around a polaron cell where the light emitted through the top p-type GaAs cap layer was collected into an optical fibre led to an Ocean Optics spectrometer. The pulse duration (1 ms) and current (0.5-80 mA) were kept minimal to avoid formation of bubbles of hydrogen in the cell which prevent current transport. Also, the LCEL measurements under the above excitation parameters caused just minimal surface damage to the cap layer and did not affect the subsequent device fabrication process.

The LCEL spectra of InP and InAsP QD samples are presented in figure 2. The emission spectrum of nominally binary InP QDs was dominated by an emission centred at 716 nm with a full width at half maximum (FWHM) of 20.6 nm (equivalent to ~50 meV). Adding arsine to the reactor during the QD growth resulted into a wavelength shift of the QD emission to 775 nm, a decreased peak intensity and an increased FWHM of 34.3 nm (~71 meV). The latter indicates a higher degree of compositional and size inhomogeneity of InAsP QDs. It is important to mention that the long wavelength shift of ~60 nm (132 meV) of the InAsP QDs is considerably smaller than expected from the arsenic fraction derived from the X-ray diffraction data and the corresponding bandgap reduction of InAsP. This discrepancy can be explained by a smaller average InAsP QD size in comparison to InP QDs, as suggested by Vinokurov *et al* [30].

4. Transmission electron microscopy (TEM)

We performed TEM studies on the QD laser samples to confirm the QD formation and assess the QD size and density. Representative TEM images are shown in figure 3. In both samples, the QDs are aligned in isolated vertical stacks with a density of the order of 10^9-10^{10} cm^{-2}. The InP QDs generally appear larger and the stacks more regular than the InAsP QDs, with a height of 4-5 nm and a lowermost dot diameter of 40-50 nm, whereas the InAsP dots have a height of 2-3 nm, a lowermost dot diameter of 20-30 nm and greater variability in dot diameter through the stack. It is not yet clear whether the smaller dot size and greater size variation is a fundamental property of the InAsP dots (e.g. due to a higher lattice mismatch) or simply results from using growth conditions optimised for InP dots, which may not be optimal for the InAsP dots. However, the observed greater variability and smaller average size of the InAsP QDs are in a good agreement with their luminescence properties described in the previous section.

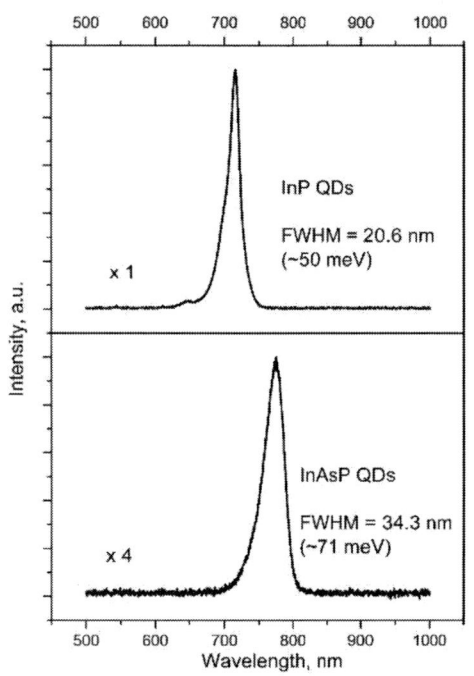

Figure 2. LCEL spectra of as grown InP and InAsP QD laser wafers at RT.

Figure 3. Transmission electron microscope images (bright field, g = 004) of InAsP QD (upper image) and InP QD (lower image) laser structures.

5. Laser results

The epitaxial wafers were processed in 4 mm long, 50 μm wide, oxide isolated stripe lasers with uncoated facets. The lasers were operated in pulsed mode (1 kHz, 1000 ns) to reduce self-heating. Laser oscillations of InP and InAsP QD samples were observed at around 720 nm and 770 nm (figure 4), respectively, close to the corresponding LCEL wavelengths. Both samples delivered optical powers of at least 200 mW.

The results of detailed measurements of the temperature dependence of the threshold current densities are presented in figure 5. They confirmed higher threshold current densities for InAsP QD samples over the studied temperature interval. However, the both structures showed reasonably small threshold current densities (e.g. 123 A/cm^2 and 155 A/cm^2 at RT for InP and InAsP QD samples, respectively) and lased up to at least 380 K. At elevated temperatures, above ~350 K, the threshold current density of the InAsP QD laser exhibits a trend of a slightly slower increase as compared to that of the InP QD laser due to the greater confinement potential in InAsP QDs. In general, the performances of InP and InAsP QD lasers compare well with those of uncoated and unmounted lasers based on other QD and quantum well materials systems operating around the same spectral range of ~700-800 nm [39, 40, 41, 42].

Detailed characterization and analysis of the laser properties of InAsP QDs can be found elsewhere [43].

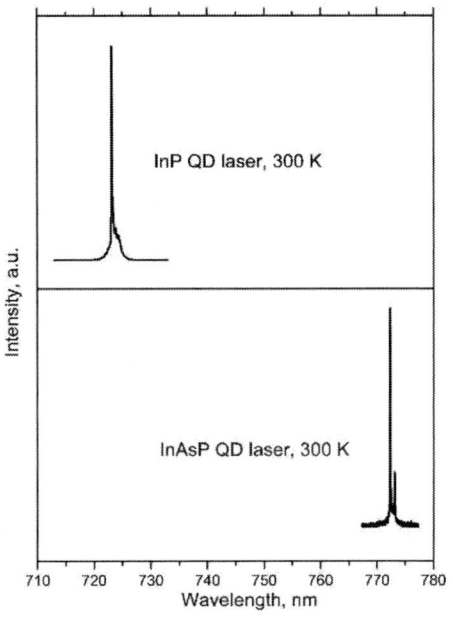

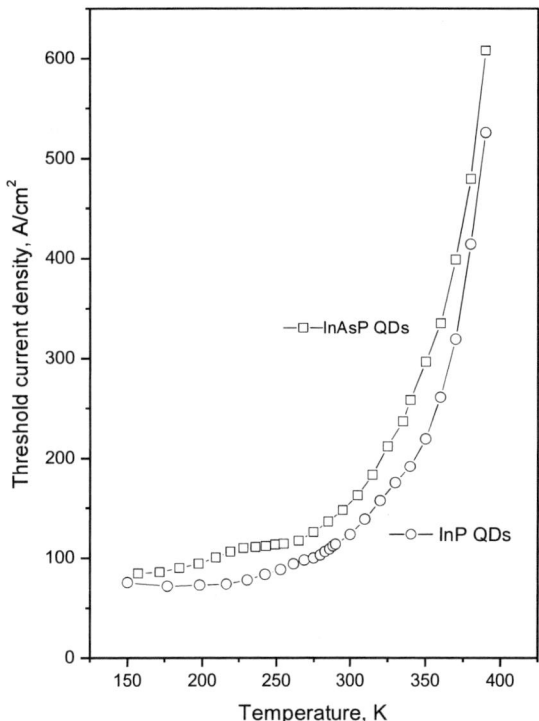

Figure 4. Emission spectra of InP and InAsP QD laser diodes just above the threshold.

Figure 5. Temperature dependence of the threshold current densities of InP and InAsP QD laser diodes.

6. Conclusions

We have demonstrated MOVPE of InAsP QDs as a feasible approach to extend the operation range of InP QD lasers towards longer wavelengths. A long wavelength shift of the QD spontaneous emission of ~60 nm was achieved at a molar As fraction of ~25% in the group V sublattice. In comparison to InP QDs, InAsP QDs shows a greater size and composition variability as evident from TEM images and increased inhomogeneous broadening of the LCEL spectra. Uncoated, 4mm long and 50 µm wide stripe InAsP QD laser operated at ~770 nm in a pulse regime with a threshold current of 155 A/cm^2 and maximum optical powers of at least 200 mW at RT and maximum operation temperature of at least 380 K.

Acknowledgments

We appreciate the Engineering and Physical Sciences Research Council (EPSRC) for funding this research (grant EP/L005409/1), and we are grateful to Mr David Morris, Chief Research Technician at the EPSRC National Centre for III-V Technologies, for his invaluable assistance around the MOVPE lab. One of the authors (ABK) appreciates the support from the Competitiveness Programme of National Research Nuclear University MEPhI.

The data associated with this paper is available from the following address: http://dx.doi.org/10.17035/d.2015.100105.

References

[1] Ahopelto J, Yamaguchi A, Nishi K, Usui A and Sakaki H 1993 Nanoscale InP Islands for Quantum Box Structures by Hydride Vapor Phase Epitaxy *Japan. J. Appl. Phys.* **32** L32–4

[2] DenBaars S P, Reaves C M, Bressler-Hill V, Varma S, Weinberg W H and Petroff P M 1994 Formation of coherently strained self-assembled InP quantum islands on InGaP/GaAs (001) *J. Cryst. Growth* **145** 721–7

[3] Carlsson N, Seifert W, Petersson A, Castrillo P, Pistol M E Samuelson L 1994 Study of the two-dimensional–three-dimensional growth mode transition in metalorganic vapor phase epitaxy of GaInP/InP quantum-sized structures *Appl. Phys. Lett.* **65** 3093–5

[4] Zundel M K, Jin-Phillipp N Y, Phillipp F, Eberl K, Riedl T, Fehrenbacher E and Hangleiter A 1998 Red-light-emitting injection laser based on InP/GaInP self-assembled quantum dots *Appl. Phys. Lett.* **73** 1784–6

[5] Porsche J, Ost M, Scholz F, Fantini A, Phillipp F, Riedl T and Hangleiter A 2000 Growth of Self-Assembled InP Quantum Islands for Red-Light–Emitting Injection Lasers, *IEEE J. Sel. Top. Quantum Electron.* **6** 482–90

[6] Walter G, Elkow J, Holonyak N, Heller R D, Zhang X B and Dupuis R D 2004 Visible spectrum (645 nm) transverse electric field laser operation of InP quantum dots coupled to tensile strained $In_{0.46}Ga_{0.54}P$ quantum wells *Appl. Phys. Lett.* **84** 666–8

[7] Smowton P M, Lutti J, Lewis G M, Krysa A B, Roberts J S and Houston P A 2005 InP-GaInP quantum-dot lasers emitting between 690-750 nm *IEEE J. Sel. Top. Quantum Electron.* **11** 1035–40

[8] Elliott S N, Smowton P M, Krysa A B and Beanland R 2012 The effect of strained confinement layers in InP self-assembled quantum dot material *Semicond. Sci. Technol.* **27** (2012) 094008

[9] Kasim M, Elliott S N, Krysa A B and Smowton P M 2015 Reducing Thermal Carrier Spreading in InP Quantum Dot Lasers *IEEE J. Sel. Top. Quantum Electron.* **21** 1900306

[10] Schlosser P J, Hastie J E, Calvez S, Krysa A B and Dawson M D 2009 InP/AlGaInP quantum dot semiconductor disk lasers for CW TEM00 emission at 716-755 nm *Optics Express* **17** 21782–7

[11] Schwarzbäck T, Bek R, Hargart F, Kessler C A, Kahle H, Koroknay E, Jetter M and Michler P 2013 High-power InP quantum dot based semiconductor disk laser exceeding 1.3 W *Appl. Phys. Lett.* **102** 092101

[12] Shutts S, Smowton P M and Krysa A B 2014 Dual-wavelength InP quantum dot lasers *Appl. Phys. Lett.* **104** 4883857

[13] Shutts S, Elliott S N, Smowton P M and Krysa A B 2015 Exploring the wavelength range of InP/AlGaInP QDs and application to dual-state lasing *Semicond. Sci. Technol.* **30** 044002

[14] Langbein W, Cesari V, Masia F, Krysa A B, Borri P and Smowton P M 2010 Ultrafast gain dynamics in InP quantum-dot optical amplifiers *Appl. Phys. Lett.* **97** 211103

[15] Savitski V G, Schlosser P J, Hastie J E, Krysa A B, Roberts J S, Dawson M D and Burns D 2010 Passive mode-locking of a Ti:Sapphire laser by InGaP quantum-dot saturable absorber *IEEE Photon. Technol. Lett.* **22** 209–11

[16] Butkus M *et al* 2011 High repetition rate Ti:sapphire laser mode-locked by InP quantum-dot saturable absorber *IEEE Photon. Technol. Lett.* **23** 1603–5

[17] Eichfelder M, Schulz W M, Reischle M, Wiesner M, Rossbach R, Jetter M and Michler P 2009 Room-temperature lasing of electrically pumped red-emitting $InP/(Al_{0.20}Ga_{0.80})_{0.51}In_{0.49}P$ quantum dots embedded in a vertical microcavity *Appl. Phys. Lett.* **95** 131107

[18] Beirne G J, Michler P, Jetter M and Schweizer H 2005 Single-photon emission from a type-B InP/GaInP quantum dot *J. Appl. Phys.* **98** 093522

[19] Reischle M, Beirne G J, Schulz W M, Eichfelder M, Roßbach R, Jetter M and Michler P 2008 Electrically pumped single-photon emission in the visible spectral range up to 80 K *Optics Express* **16** 12771–6

[20] Roßbach R, Schulz W M, Reischle M, Beirne G J, Jetter M and Michler P 2008 Increased single-photon emission from InP/AlGaInP quantum dots grown on AlGaAs distributed Bragg reflectors *J. Cryst. Growth* **310** 4818–20

[21] Schulz W M, Roßbach R, Reischle M, Beirne G J, Bommer M, Jetter M and Michler P 2009 Optical and structural properties of InP quantum dots embedded in $(Al_xGa_{1-x})_{0.51}In_{0.49}$ *Phys. Rev. B* **79** 035329

[22] Skiba-Szymanska J, Chekhovich E A, Nikolaenko A E, Tartakovskii A I, Makhonin M N, Drouzas I, Skolnick M S and Krysa A B 2008 Overhauser effect in individual InP/Ga_xIn_{1-x}P dots *Phys. Rev. B* **77** 165338

[23] Chekhovich E A, Makhonin M N, Kavokin K V, Krysa A B, Skolnick M S and Tartakovskii A I 2010 Pumping of nuclear spins by optical excitation of spin-forbidden transitions in a quantum dot *Phys. Rev. Lett.* **104** 066804

[24] Chekhovich E A, Krysa A B, Skolnick M S and Tartakovskii A I 2011 Direct measurement of the hole-nuclear spin interaction in single InP/GaInP quantum dots using photoluminescence spectroscopy *Phys. Rev. Lett.* **106** 027402

[25] Chekhovich E A, Kavokin K V, Puebla J, Krysa A B, Hopkinson M, Andreev A D, Sanchez A M, Beanland R, Skolnick M S and Tartakovskii A I 2012 Structural analysis of strained quantum dots using nuclear magnetic resonance *Nature Nanotechnology* **7** 646–50

[26] Chekhovich E A, Glazov M M, Krysa A B, Hopkinson M, Senellart P, Lemaître A, Skolnick M S and Tartakovskii A I 2013 Element-sensitive measurement of the hole-nuclear spin interaction in quantum dots *Nature Physics* **9** 74–8

[27] Chekhovich E A, Makhonin M N, Skiba-Szymanska J, Krysa A B, Kulakovskii V D, Skolnick M S and Tartakovskii A I 2010 Dynamics of optically induced nuclear spin polarization in individual InP/Ga_xIn_{1-x}P quantum dots *Phys. Rev. B* **81** 245308.

[28] Luxmoore I J, Ahmadi E D, Wasley N A, Fox A M, Tartakovskii A I, Krysa A B and Skolnick M S 2010 Control of spontaneous emission from InP single quantum dots in GaInP photonic crystal nanocavities *Appl. Phys. Lett.* **97** 181104

[29] Qiu Y, Krysa A B and Walther T 2010 STEM imaging of InP/AlGaInP quantum dots *J. Phys. Conf. Ser.* **245** 012087

[30] Vinokurov D A, Kapitonov V A, Kovalenkov O V, Livshits D A and Tarasov I S 1998 Self-organized nanosize InP and InAsP clusters obtained by metalorganic compound hydride epitaxy *Technical Physics Letters* **24** 623–5

[31] Fuchi S, Miyake S, Kawamura S, Lee W S, Ujihara T and Takeda Y 2008 Effects of absorbed group-V atoms on the size distribution and optical properties of InAsP quantum dots fabricated by the droplet hetero-epitaxy *J. Cryst. Growth* **310** 2239–43

[32] Ohba Y, Ishikawa M, Sugawara H, Yamamoto M and Nakanisi T 1986 Growth of high-quality InGaAlP epilayers by MOCVD using methyl metalorganics and their application to visible semiconductor lasers *J. Cryst. Growth* **77** 374–9

[33] Hamada H, Shono M, Honda S, Hiroyama R, Yodishi K and Yamaguchi T 1991 AlGaInP visible laser diodes grown on misoriented substrates *IEEE J. Quantum Electron.* **27** 1483–90

[34] Zorn M, Wenzel H, Zeimer U, Sumpf B, Erbert G and Weyers M 2007 High-power red laser diodes grown by MOVPE", *J. Cryst. Growth* **298** 667–71

[35] Zory P S, Young C L, Hsu C F, O J S, Largent C C 1995 Diode Laser Material Evaluation Using Liquid Contact Luminescence *8th LEOS Ann. Meeting,* 30 Oct. - 2 Nov. San Francisco, Conf. Proc., vol. 2., pp. 133–4 http://dx.doi.org/10.1109/LEOS.1995.484632

[36] Hsu C F, Largent C C, O J S, Young C L, Zory P S and Bour D P 1996 Internal Quantum

Efficiency Measurements of GaInP Quantum Well Laser Material Using Liquid Contact Luminescence *Proc. SPIE* **2682** 136–42 http://dx.doi.org/10.1117/12.237650

[37] Largent C C, Zory P S and Bour D P 1997 Liquid contact luminescence for laser material evaluation and flat panel display", *10ᵗʰ LEOS Ann. Meeting, 10-13 Nov. San Francisco, Conf. Proc.* vol 2, pp. 107–8 http://dx.doi.org/10.1109/LEOS.1997.645284

[38] Krysa A B, Liew S L, Lin J C, Roberts J S, Lutti J, Lewis G M and Smowton P M 2007 Low threshold InP/AlGaInP on GaAs QD laser emitting at ~740 nm *J. Cryst. Growth* **298** 663–6

[39] Schlereth T W, Gerhard S, Kaiser W, Höfling S and Forchel A 2007 High-performance short-wavelength (~760 nm) AlGaInAs Quantum-Dot Lasers *IEEE Photon. Technol. Lett.* **19** 1380–2

[40] Erbert G, Bugge F, Knauer A, Sebastian J, Thies A, Wenzel H, Weyers M and Trankle G 1999 High-power tensile-strained GaAsP–AlGaAs quantum-well lasers emitting between 715 and 790nm *IEEE J. Sel. Top. Quantum Electron.* **5** 780–4

[41] Agahi F, Lau K M, Choi H K, Baliga A and Anderson N G 1995 High-performance 770-nm AlGaAs–GaAsP tensile-strained quantum-well laser-diodes *IEEE Photon. Technol. Lett.* **7** 140–3

[42] Mawst L J, Rusli S, Al-Muhanna A and Wade J K 1999 Short-wavelength (0.7 μm < λ < 0.78 μm) high-power InGaAsP-active diode lasers *IEEE J. Sel. Topics Quantum Electron.* **5** 785–91

[43] Karomi I, Smowton P M, Shutts S, Krysa A B and Beanland R 2015 InAsP Quantum Dot Lasers Grown by MOVPE *Optics Express* **23** 27282–91

CORSCS2015   IOP Publishing
Journal of Physics: Conference Series **740** (2016) 012009   doi:10.1088/1742-6596/740/1/012009

High quality $Y_3Al_5O_{12}$ doped transparent ceramics for laser applications, role of sintering additives

A A Kaminskii[1], V V Balashov[2], E A Cheshev[3,4], Yu L Kopylov[2], A L Koromyslov[3], O N Krokhin[3,4], V B Kravchenko[2], K V Lopukhin[2], V V Shemet[2] and I M Tupitsyn[4]

[1] Shubnikov Crystalography Institute RAS, 59 Leninsky pr., 119333, Moscow, Russia
[2] Kotelnikov FIRE RAS, 1 Vvedensky Sq., 141120 Fryazino, Russia
[3] P.N. Lebedev Physical Institute of the RAS, 53Leninsky pr., 119991, Moscow, Russia
[4] National Research Nuclear University MEPhI, 31 Kashirskoye shosse, 115409, Moscow, Russia

E-mail: ylk215@yandex.ru, additional ylk215@ire216.msk.su

Abstract. SiO_2, ZrO_2, B_2O_3 and MgO oxides and their combinations were used as sintering aids for preparation of yttrium aluminum garnet (YAG) ceramics doped by Nd_2O_3, Er_2O_3, Ho_2O_3, Tm_2O_3 and Yb_2O_3. The influence of these additives on optimal sintering temperature, grain growth, volume of residual pores and optical quality of the ceramics were investigated. The best combination of the sintering additives was found and high quality samples of YAG:Nd (1 at.%) ceramics were obtained. The original method of laser optical quality characterization of ceramics was developed and tested. The main laser parameters of YAG:Nd (1 at.%) ceramics samples are measured and compared with the best well known laser ceramics. The samples of YAG:RE (RE- Er_2O_3, Ho_2O_3, Tm_2O_3 and Yb_2O_3) ceramics are obtained, and their optical transmittance spectra are measured. Composite structures of YAG:Yb (5 at.%) − YAG were obtained by the simplest method of successive joint compaction of different composition layers.

1. Introduction

The main problem of doped YAG ceramics production technology with laser level of samples quality is elimination of residual pores. In the frames of solid state reactive sintering process a large number of factors effects on residual porosity such as appropriate morphology and dispersivity of starting oxide powders [1-4], stoichiometry of composition [5], conditions of compaction and sintering [6-10] and the presence of sintering additives (SA). There are attempts to produce ceramics without any (special) SA (see, for example, [11]). But practically all high quality ceramics were obtained with SA. As a rule, SA for oxide ceramics are different oxides. The most traditional SA for YAG is SiO_2 [12-15]. MgO and SiO_2 and MgO combinations were used in some very interesting and successful works [16-18]. There is no significant difference in results with SiO_2 and MgO, but in case of B_2O_3 and

---

Content from this work may be used under the terms of the Creative Commons Attribution 3.0 licence. Any further distribution of this work must maintain attribution to the author(s) and the title of the work, journal citation and DOI.
Published under licence by IOP Publishing Ltd

combination of B_2O_3 and SiO_2 as SA there is a sharp contrast [19]. It was interesting to expand the research of application of the individual oxides and their possible combinations as sintering additives.

2. Experimental

High purity oxides Y_2O_3, Nd_2O_3, Er_2O_3, Ho_2O_3, Tm_2O_3 and Yb_2O_3 produced by Lanhit Ltd. and Al_2O_3 - AKP- 50 produced by Sumitomo Chem. Corp. and BMA 15 produced by Baikowski Corp. were used as starting powder materials for ceramics samples preparation. SiO_2, ZrO_2, B_2O_3 and MgO in different combinations were used as SA for $(Y_{1-x}REx)_3Al_5O_{12}$ (YAG:RE) compositions. The concentration ranges of SA were the following: B_2O_3–(0.45 -1.5) mol.%; SiO_2–(0.45–1.35) mol. %; MgO-(0.05–0.45) mol. %; ZrO_2- 0.2 mol.%. All additives were reagent grade. Powders were weighed in ratios corresponding to the chemical composition and were mixed in planetary ball mill with anhydrous isopropanol and alumina balls. After milling and drying, the powders were sieved through 200 mesh sieve and mixed in the ball mill again with anhydrous isopropanol and PVB. After repeated drying and sieving the powders were pressed uniaxially at 50 MPa. PVB was evaporated at 800°C and compacts were finally pressed isostatically at 200-250 MPa. Compacts of YAG:RE compositions were vacuum sintered at 1500-1780°C. The heating rate was about 0,3°C/min at the temperatures around the maximum rate of shrinkage. Ceramics samples after sintering were annealed in air at 1100°C during 32 h. For crystallographic phase characterization the X-ray diffraction (XRD) machine D8 DISCOVER (Cu Kα1,2 λ = 1,542Å) was used. Shrinkage investigations during the sintering – thermomechanical analysis (TMA) were conducted with DIL402C, Netzsch and thermogravimetric and scanning calorimetric analysis (TGA, DSC) with NETZSCH STA 449 machines respectively. For pores contents measurement the light tomography method was used similar to the method described in [20]. The special setup was developed and used for lasing experiments. The description of this method is presented below.

3. Results and discussion

The temperature range of maximum rate of shrinkage during the sintering process for YAG: RE compositions depends strongly on the type and the combination of SA. The most remarkable combinations of SA investigated can be divided in to three groups: "1"- B_2O_3 (0.9)+SiO_2(0.9); "2"- B_2O_3(1,5)+SiO_2(0.9)+ZrO_2(0.2); "3"- SiO_2(1.35)+MgO(x). The results of TMA investigation for these groups are shown in figure 1.

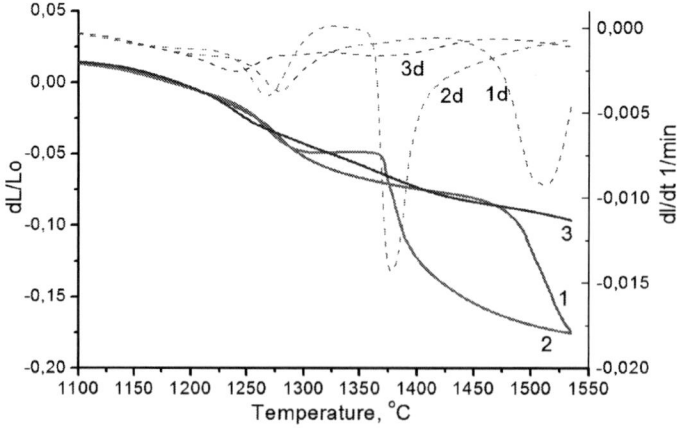

Figure 1. Temperature dependence of shrinkage (1,2,3) and rate of shrinkage (1d,2d,3d) for three groups of YAG: Nd (1at.%) ceramics.

Generally, the increase the SA concentration, especially B_2O_3, shifts the maximum of shrinkage rate to lower temperatures. It was found, that it is possible to increase significantly B_2O_3 and SiO_2 contents without the deterioration of optical quality of the ceramic samples. At the same time the increase of ZrO_2 and MgO contents for more than 0.2 mol % results in increasing of residual porosity, and as a result the optical transmittance of ceramics drops down dramatically. XRD diagrams show that for B_2O_3 and/or SiO_2 contents above 0.6 mol % (that is for groups 1 and 2 in figure 1) the formation of garnet structure during the sintering was completed below 1550°C. In the contrast about 7 % of $Y_2Al_4O_9$ and $YAlO_3$ phases remained for group 3 at 1500°C temperature, and the final formation of YAG phase was completed after 1550-1600°C. The difference of aluminate phases formation in groups 1 and 2 are demonstrated by general XRD diagrams presented in figure 2.

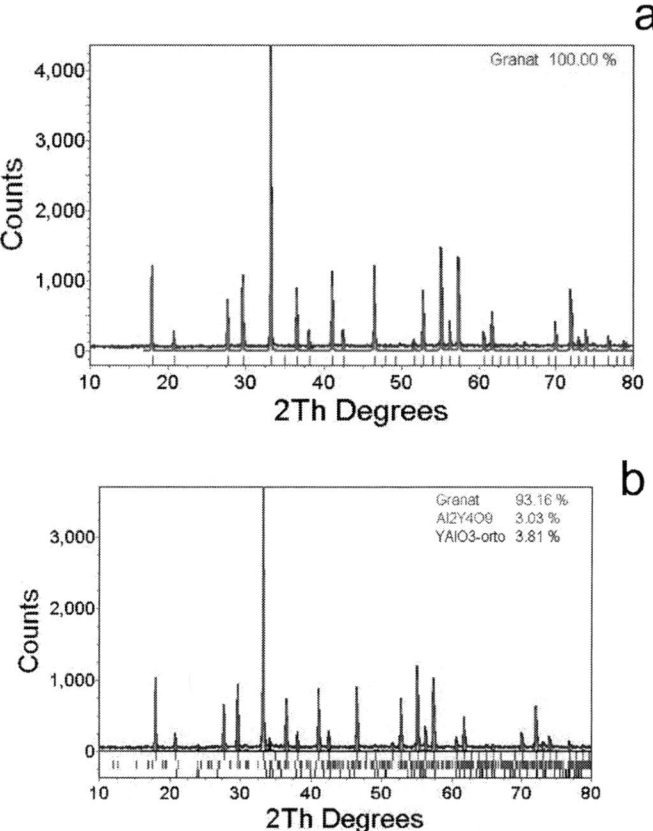

Figure 2. XRD diagrams of YAG samples sintered in dilatometer at 1550°C . Samples of 1 and 3 groups (a). samples of group1(b).

In shrinkage curves for all samples with content B_2O_3 there is a temperature range where shrinkage rate is near to zero. It is most pronounced for samples of group "2". It can be explained by the reorientations of powder particles in presence of liquid phase created by SA. As was shown in [20-22] this reorientation range in presence of liquid phase comes really before shrinkage. The shift of maximum rate of shrinkage to lower temperatures increases the grain size and optical transmittance at every sintering temperature and decreases the residual porosity, as shown in figure 3 a, b and c.

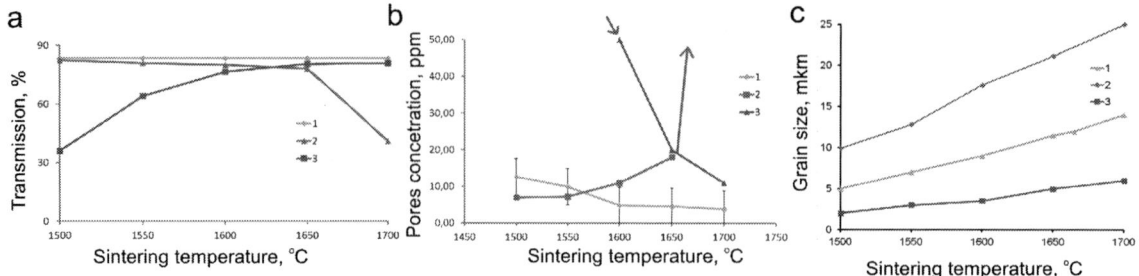

Figure 3. Characteristics of three groups (1,2,3) ceramic samples vs sintering temperature. a) Optical transmittance at λ-1063 nm. b) Residual pores concentration. c) Grain size.

The optical transmission reaches almost the theoretical level for group 1 before 1500°C and for group 3 before 1700°C. For group 2 transmittance drops down near 1700°C (see figure 3 a). These data are in reasonable correlation with concentration of the residual porosity (see figure 3 b). Unfortunately, at the same time the optical transmittance is growing together with increasing of grain size (compare the ratio between transmittance, porosity, figure 3 a, b) and grain size figure 3 c)). When a compromise between the grain size and optical transparency is found, it is possible to obtain the ceramics with high optical transparency for YAG: Nd (or Yb) samples with appropriate combination of SA. Examples of such ceramics and spectra of optical transmittance are presented in figure 4 for samples of YAG:Nd (1at.%).

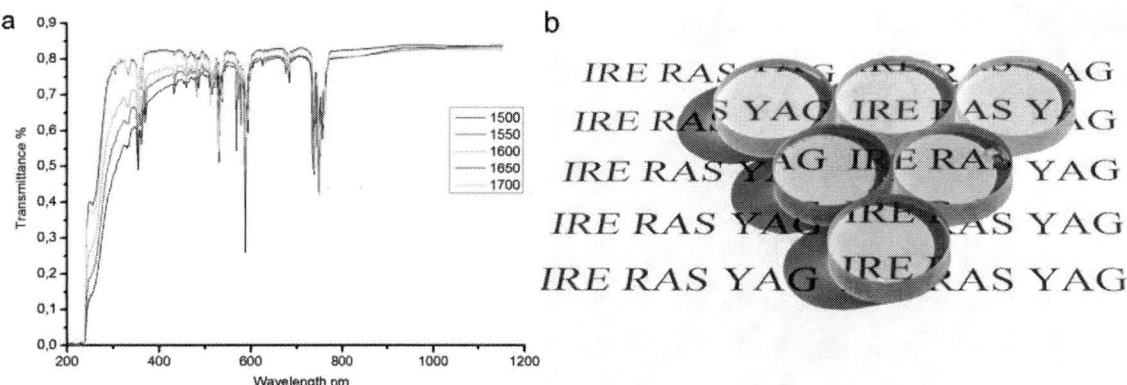

Figure 4. Optical inline transmittance of ceramics samples from group "2" (a) and general view of these ceramics (b). Insertion in a) is the sintering temperatures. Diameter of samples in b) is equal to 21 mm.

Dramatical decrease of transmittance and increase of the residual porosity for group 2 of samples at the sintering temperature can be explained probably by the presence of heterovalent zirconium cations.

The original method for measurement of lasing threshold at the longitudinal diode pumping nearly transverse mode locking was used for testing of lasing properties of the produced samples. The details of this method are described in detail elsewhere [23-25]. For the measurement procedure, the cavity length L of resonator is changed and power of the lasing threshold is automatically measured. The pumping of the sample is not uniform, and the transverse mode locking takes place at some values of length, L. Conditions of transverse mode locking can be determined by the following equation–condition of transverse mode locking for empty cavity [26]:

$$\arccos(\sqrt{g_1 g_2}) = \pi \frac{r}{s},$$

where r/s – proper fraction which characterize the degeneration [26], $g_{1,2} = 1 - L/R_{1,2}$ cavity stability parameter, L – cavity length, R – radius of mirrors.

At every L, near transverse mode locking, the pumping power value of lasing threshold decreases dramatically, and the value of threshold power at these conditions is very sensitive to any small optical heterogeneity of the laser media. Using the Konoshima Chemical corp. ceramic sample as etalon the spectra of pumping power value of lasing threshold were measured for four samples of our YAG: Nd (1at. %) ceramics with different concentration of residual pores. Results of these measurements are presented in figure 5.

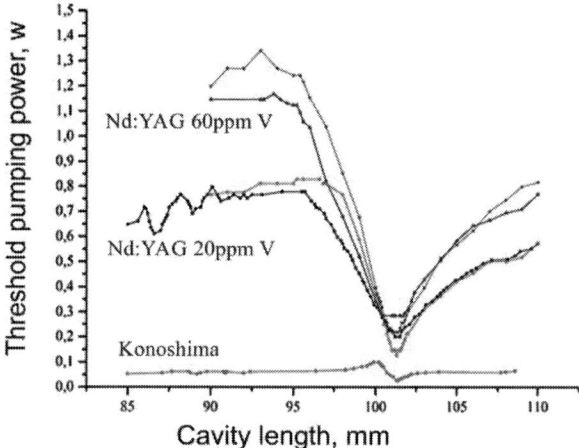

Figure 5. The pumping power corresponding to lasing threshold as a function of cavity length near the L value, where the proper fraction r/s is equal to 1/4 for Konoshima ceramic YAG:Nd (1 at.%) sample and for two groups of ceramic YAG:Nd (1 at.%) samples with 20 and 60 ppm pores volume fractions.

The presented results show clearly that this method is really sensitive for defect detection in ceramic samples especially in the case of small concentration of residual pores. These data correspond to results obtained in work [27] by direct measuring output power of laser with ceramic samples which have different residual porosity. As it is seen in figure 5 the increasing of volume pores concentration from 1-2 ppm (for Konoshima ceramics) to 20 ppm and from 20 to 60 ppm increases the lasing threshold power by 10 and 20 times respectively. The comparative spectra for Konoshima's samples and our high quality samples like shown in figure 4 are presented in figure 6.

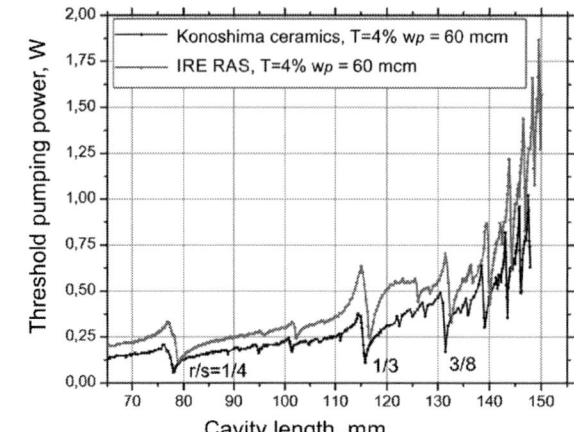

Figure 6. Pumping power corresponding to the lasing threshold as a function of cavity length for two YAG:Nd(1at. %) ceramic samples measured at the same conditions. w_ρ is radius of pumping beam, T=4% is reflectivity of output mirror.

Results for both types of samples are very close each to other. Some main laser characteristics were also measured for these high quality samples, and the results of these measurements are shown in figure 7. The data in figure 7 a) were obtained at the conditions when the pumping power was absorbed completely along the sample length. The slope efficiency determined from these measurements is 72 and 55 % for Konoshima samples and our ceramics respectively. The pulse duration is identical for both ceramics samples (see figure 7 b). The same quality ratio between the tested ceramic samples was found for such parameters as pulse power (figure 7 d), average output power (figure 7 e) and pulse duration (figure 7 c) as a function of output mirror reflectivity.

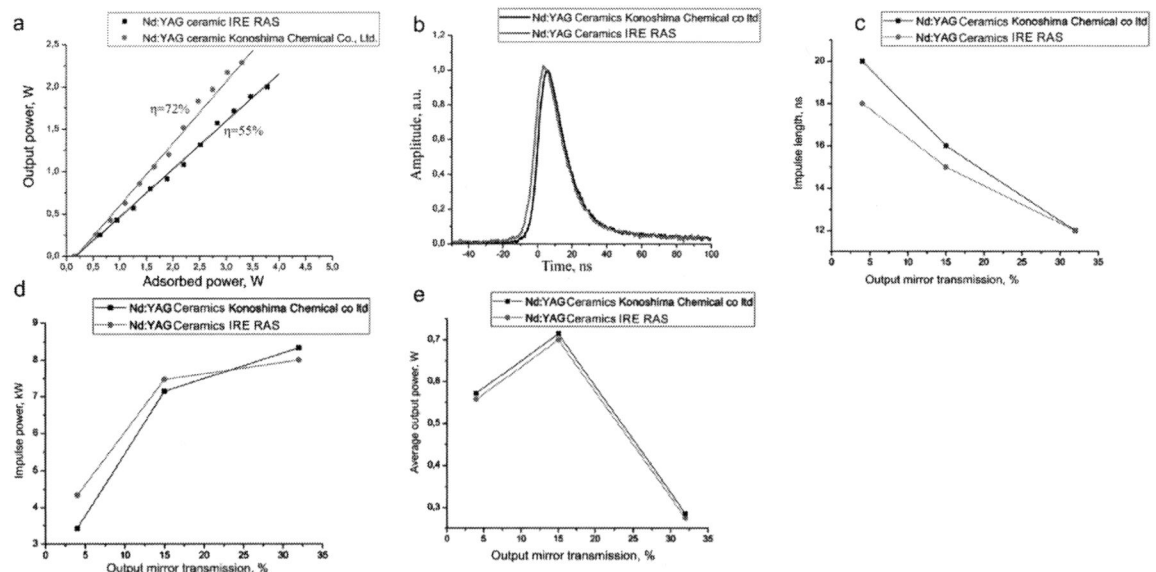

Figure 7. The main lasing characteristics of produced YAG:Nd(1at.%) ceramics in comparison with Konoshima etalon. Output power vs input power (a); impulse duration and shape (b), impulse length (c), impulse power (d) and average output power (e) as a function of output mirror transmission.

BMA-15 powder was used as Al_2O_3 source for manufacturing of samples YAG:RE, where RE=Yb_2O_3, Er_2O_3, Ho_2O_3, Tm_2O_3. Unfortunately, for samples of YAG with BMA-15 we did not achieve the same optical quality as in the case of AKP-50 powder. In these samples the residual pores in concentration about 20 ppm are existing. Probably it is due to not optimal milling condition for this powder. In figure 8 a, b the optical transmittance spectra and general view of $(Y_{2.91}\ Tm_{0.09})Al_5O_{12}$; $(Y_{2.85}\ Yb_{0.15})Al_5O_{12}$; $(Y_{2.985}\ Ho_{0.015})Al_5O_{12}$; $(Y_{2.985}\ Er_{0.015})Al_5O_{12}$ ceramics samples are presented.

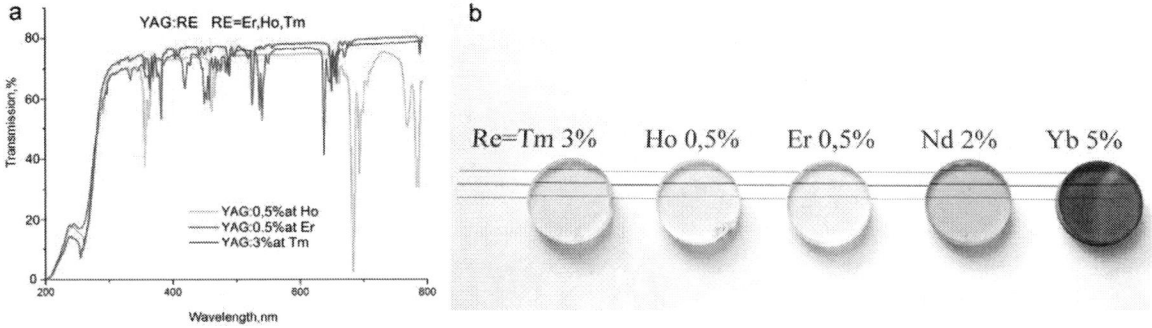

Figure 8. Optical in-line transmittance of $(Y_{2.91}\ Tm_{0.09})Al_5O_{12}$; $(Y_{2.85}\ Yb_{0.15})Al_5O_{12}$; $(Y_{2.985}\ Ho_{0.015})Al_5O_{12}$; $(Y_{2.985}\ Er_{0.015})Al_5O_{12}$ ceramics samples (a) and general view of the samples (b). Diameter of all samples in the photo is 21mm.

For YAG doped by Yb^{3+} the most attractive design of laser active elements is disk type. In this case, because the disc is thin enough, the composite structure with thick passive part is most profitable. Composite element must combine in monolithic ceramic body active (doped) and not active (not doped) parts. The simple method was used for preparation of the composite. In the metallic die the layer of YAG powder doped with Yb was pressed at 15-20 MPa, then the die was filled by non-doped YAG powder and pressed again at 50 MPa. The preformed compact was CIPed at 230 MPa and sintered by the method described early. The general view of the produced composite is shown in figure 9 a and b. After vacuum sintering Yb in YAG is in Yb^{2+} valence and has a green color. It is very comfortable for measurements of Yb^{2+} ions distribution by probe laser beam (λ-630 nm) scanning.

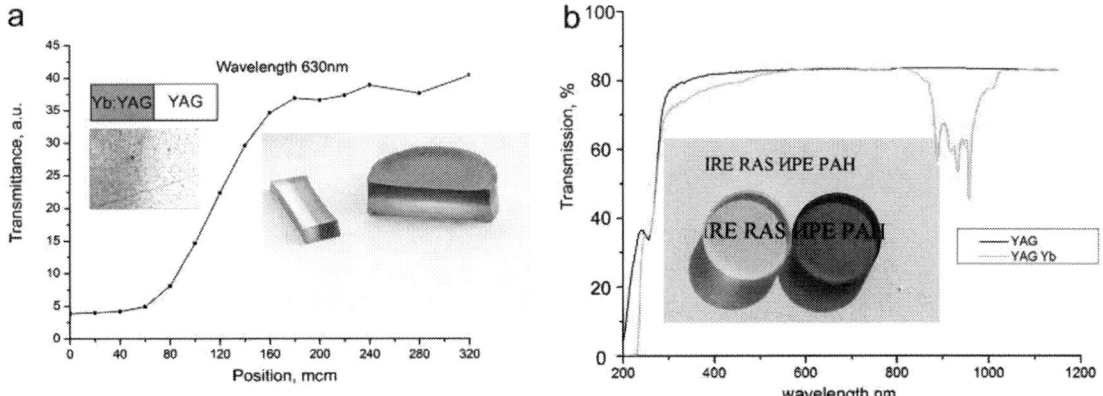

Figure 9. YAG:Yb(5 at.%) / YAG composite ceramics. In both photos in inserts (a, b) ceramic samples are as vacuum sintered and therefore Yb is as Yb^{2+}, that is green color. Distribution of Yb^{2+} near the border of two parts of the composite (a) which is measured as transmittance of light at λ-630 nm. Optical transmittance of doped and un-doped parts of the composite (b).

The results of these scanning measurements are shown in figure 9 b. It was found from these measurements that the width of transient region between two parts of the composite is about a hundred microns. It is a big value, and probably it can be explained by availability of intergrain fast diffusion. The optical transmission spectra of doped and undoped parts of YAG:Yb (5 at.%) - YAG composite are presented.

4. Conclusions

The role of sintering additives chosen from the group of SiO_2, ZrO_2, B_2O_3 and Mg in high quality YAG ceramics production was investigated. When contents of SiO_2 and especially B_2O_3 is growing up the grain size is increased, but the residual porosity decreased and the sintering temperature at which the optical transmittance achieves the theoretical values also decreased. The compromise between the grain size and residual porosity (optical transmittance) was found. The original method for estimation of ceramics optical quality was developed and tested on ceramics samples. Some of the main laser characteristics of produced ceramics YAG:Nd(1 at.%) were measured and compared with the best-known laser ceramics. The quality of the produced ceramics is comparable with the etalon. Composite structures of YAG:Yb (5 at.%) – YAG were obtained, and it was shown that consistent compaction of layers of different composition can be used as simple method for composite structures formation.

Acknowledgment

This work was performed in National Research Nuclear University MEPhI and in P.N. Lebedev Physical Institute of the Russian Academy of Sciences under the Agreement № 14.575.21.0047 with Ministry of Education and Science of Russian Federation, the unique identification number of applied scientific research RFMEFI57514X0047.

The authors acknowledge the generous financial support from Program of Presidium of Russian Academy of Sciences № P25, and Russian Foundation for Basic Research, project № 14- 02- 90446 Укр_а.

This work was supported by the Competitiveness Program of National Research Nuclear University MEPhI.

References

[1]   Esposito L, Costa A L and Medri V 2008 Reactive sintering of YAG-based materials using micrometer-sized powders *J. Europ.Ceram. Soc.* **28** 1065-71

[2]   Ikesue A, Furusato I and Kamata K 1995 Fabrication of polycrystalline, transparent YAG ceramics by a solid-state reaction method *J. Am. Ceram. Soc*. **78**(1) 225

[3]   Liu Jun *et al* 2014 Effects of ball milling time on microstructure evolution and optical transparency of Nd:YAG ceramics *Ceram. Int.* **40** 7(A) 1271-8

[4]   Li Xiaodong, Li Ji-Guang, Xiu Zhimeng, Huo Di and Sun Xudong 2009 Transparent Nd:YAG Ceramics Fabricated Using Nanosized γ-Alumina and Yttria Powders *J. Am. Ceram. Soc.* **92** (1) 241–4

[5]   Patel A P, Levy M R, Grimes R W, Gaume R M, Feigelson R S, McClellan K J and Stanek C R 2008 Mechanism of nonstiochiometry in $Y_3Al_5O_{12}$ *Appl. Phys. Lett.* **93** 191902

[6]   Boulesteix R, Maitre A, Chretien L, Rabinovitch Y and Salle C 2013 Microstructural evolution during vacuum sintering of yttrium aluminum garnet transparent ceramics: toward the origin of residual porosity affecting the transparency *J. Am. Ceram. Soc.* **96** 1724-31

[7]   Kopylov Yu L, Kravchenko V B, Bagayev S N, Shemet V V, Komarov A A, Karban O F and Kaminskii A A 2009 Development of Nd^{3+}:$Y_3Al_5O_{12}$ Laser Ceramics by High-Pressure Colloidal Slip-Casting (HPCSC) Method *Opt.Mater.* **31** (5) 707-10

[8]   Lin Ge, Jiang Li, Zhiwei Zhou, Binglong Liu, Tengfei Xie, Jing Liu, Huamin Kou, Yun Shi, Pan Yubai and Jingkun Guo 2015 Nd:YAG transparent ceramics fabricated by direct cold isostatic pressing and vacuum sintering *Opt. Mater.* **50** Part A December pp 25–31

[9] Kwadwo A A, Messing G L and Dummc J Q 2008 Aqueous slip casting of transparent yttrium aluminum garnet (YAG) ceramics *Ceram. Int.* **34**(5) pp 1309-13

[10] Wei Zhang, Lu Tiecheng, Maa B, Wei Nian, Lu Zhongwen, Li Feng, Guan Yongbing, Chen Xingtao, Liu Wei and Qi Lu 2013 Improvement of optical properties of Nd:YAG transparent ceramics by post-annealing and post hot isostatic pressing *Opt. Mater.* **35** 2405-10

[11] Wang Z, Zhang Le, Yang H, Zhang J, Wang L, Zhang Q 2016 High optical quality Y_2O_3 transparent ceramics with fine grain size fabricated by low temperature air pre-sintering and post-HIP treatment *Ceram. Int.* **42** 4238–45

[12] Ge L, Li Jiang, Zhou Zhiwei, Liu Binglong, Xie Tengfei, Liu Jing, Kou Huamin, Shi Yun, Pan Yubai and Guo Jingkun 2011 Effect of SiO_2 on Densification and Microstructure Development in Nd:YAG Transparent Ceramics *J. Am. Ceram. Soc.* **94** (5) 1380–7

[13] Yagi H, Yanagitani T and Ueda K-I 2006 Nd3+:$Y_3Al_5O_{12}$ laser ceramics: Flashlamp pumped laser operation with a UV cut filter. *J. Alloys Compd.* 421(1–2) 195

[14] Yagi H, Yanagitani T, Takaichi K, Ueda K I and Kaminskii A A 2007 Characterizations and laser performances of highly transparent Nd^{3+}:$Y_3Al_5O_{12}$ laser ceramics *Opt. Mater.* **29** 1258-62

[15] Gaume R, Markosyan He A and Baer R L 2012 Effect of Si-induced defects on 1 µm absorption in laser-grade YAG ceramics *J.Appl.Phys.* **111** 093104

[16] Li Y K *et al* 2010 Fabrication of Nd:YAG transparent ceramics with TEOS, MgO and compound additives as sintering aids *J. Alloy. Compd.* **502**(1) 225e30

[17] Yang H *et al* 2012 The effect of MgO and SiO_2 codoping on the properties of Nd:YAG transparent ceramic *Opt. Mater.* **34**(6) 940e3

[18] Chen J C *et al* 2014 4350W quasi-continuous-wave operation of a diode face-pumped ceramic Nd:YAG slab laser *Opt. Laser. Technol.* **63** 50-3

[19] Stevenson A J, Li X, Martinez M A, Anderson J , Suchy D L, Kupp E R, Dickey E C, Mueller K T and Messing G L 2011 Low temperature, transient liquid phase sintering of B_2O_3-SiO_2-doped Nd:YAG transparent ceramics *J. Mater. Res.* **26** 2022

[20] Oh-Hun and Messsing G L 1991 A theoretical analysis of solution-precipitation controlled densification during piqued phase sintering *Acta Metall, Mater.* **39**(9) 2059-68

[21] Kingery W D 1959 Densification during sintering in the presence of a liquid phase. I. Theory *J. Appl. Phys.* **30** 301-6

[22] Kingery W D and Narasimham M D 1959 Densification during sintering in the presence of a liquid phase. II. Experimental *J. Appl. Phys.* **30** 307-10

[23] Bezotosnyi V V, Cheshev E A, Gorbunkov M V, Koromyslov A L, Kostryukov P V, Krivonos M S, Popov Y M, Tunkin V G Behavior of threshold pump power of diode end-pumped solid-state lasers in critical cavity configurations 2015 *Laser Phys. Lett.* **12** (2) 025001

[24] Bezotosnyi V V, Krokhin O N, Koromyslov A L Cheshev E A, Kopylov Y L, Kravchenko V B, Lopukhin K V, Tupizin I M 2015 Generation characteristics of YAG:Nd laser with YAG: Cr^{4+} passive lock on the base of oxide ceramics *Proceedings of V symposium on coherent optical radiation of semiconductor compounds and structures. Moscow-Zvenigorod 23-26 November* (Moscow, Lebedev Physical Institute RAS) (in Russian)

[25] Kaminskii A A, Balashov V V, Demianova L, Kopylov Y L, Kravchenko V B, Lopukhin K V, Lyapin A A, Lysenko S L, Ryabochkina P A and Shemet V V 2015 Transparent Y_2O_3 and $Y_3Al_5O_{12}$ Ceramics Doped by Rare Earth Cations - Technology, Optical Properties, Problems and Prospects *Proceedings of 11-th laser ceramics symposium LCS-2015 – International symposium on transparent ceramics for photonics applications, 30 Nov.- 4 Dec. Xuzhou, China* pp 63-4

[26] Wu H H, Sheu C C, Chen T W, Wei M D and Hsieh W F 1999 *Opt. Commun.* **165** 225-9

[27] Boulesteix R, Maître A, Baumard J-F, Rabinovitch Y and Reynaud F 2010 Light scattering by pores in transparent Nd:YAG ceramics for lasers: correlations between microstructure and optical properties *Opt. Express* **18** (14) 14992

8 nm nanodiamonds as markers for 2 photon excited luminescent microscopy

A Kharin[1], A Rogov[2], A Geloen[3], V Lysenko[4] and L Bonacina[2]

[1] National Research Nuclear University MEPhI, 31 Kashirskoye shosse, 115409, Moscow, Russia
[2] GAP-Biophotonics, University of Geneva, 22, ch. de Pinchat, CH-1211 Geneva 4, Switzerland
[3] University of Lyon, CarMeN Laboratory, UMR INSERM 1060, INSA de Lyon, University of Lyon, France
[4] University of Lyon, Nanotechnology Institute of Lyon (INL) UMR 5270, CNRS, INSA Lyon, Villeurbanne, F-69621, France

E-mail: [1]alexankhar@gmail.com

Abstract. Structural and luminescent properties of stable suspensions of fluorescent nanodiamonds were investigated. Measurement of the effective hydrodynamic radius yields particles less than 30 nm diameter, while the TEM measurements made on the same particles shows average diameter about 8 nm. It was found that NDs have relatively low toxicity. Upon incubation, 3T3-L1 cells spontaneously take up nanodiamonds that uniformly distribute in cells cytoplasm. The possibility of fluorescent imaging using both single ore two-photon excitation was shown.

1. Introduction

Carbon nanoparticles are ones of key materials for biomedical applications as they are biocompatible, biodegradable and are capable of providing novel imaging and therapeutic modalities, which are usually attributed to efficient photoluminescence either through quantum-confined states in small carbon particles or through the doping centers [1]. As it was shown before, [2, 3] nanodiamonds are one of the least toxic through all carbon nanomaterials. Nevertheless, the application of the luminescent nanodiamonds for bioimaging is limited by the low light penetration depth. Bulk diamond has bandgap 5.5 eV, so it should be excited with far-UV light. Nitrogen-dopped nanodiamonds can be excited by the significantly lower energies in UV-visible range. It is widely known, that mammalians' body has near-infrared window and the maximum penetration depth at the range 650-1350 nm while the UV and visible light is effectively absorbed by the body. One of the ways to increase the penetration depth is to excite the luminescence using the 2-photon absorption way [4]. It requires intensive laser irradiation, but the longer wavelength of the exciting irradiation can provide higher light penetration length. The up-to date studies usually use combustion-derived nanodiamonds with sizes from tenth to hundereds nanometers for single photonor multy-photon imaging [4]. Here, we by *in vitro* studies have showed that laser-ablation synthesized ultrapure nanodiamonds with sizes of 8-10 nm, could also be used as efficient non-linear optical labels with enhanced two-photon excited photoluminescence.

Content from this work may be used under the terms of the Creative Commons Attribution 3.0 licence. Any further distribution of this work must maintain attribution to the author(s) and the title of the work, journal citation and DOI.
Published under licence by IOP Publishing Ltd

2. Synthesis and characterisation of nanodiamonds

Nanodiamonds with the sizes about 8 nm were ordered at Ray Techniques Ltd. The synthesis method of such nanodiamonds is Light Hydro-dynamic Pulse (LHDP) fabrication and described in details in [5]. High resolution TEM picture in figure 1a allows to clear visualize their crystalline nature. Atomic inter-plane distances estimated from the HRTEM pictures corresponds to the very well-known diamond crystalline structure.

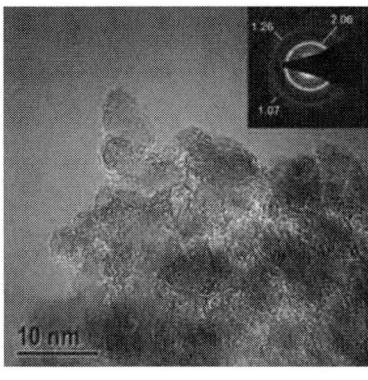
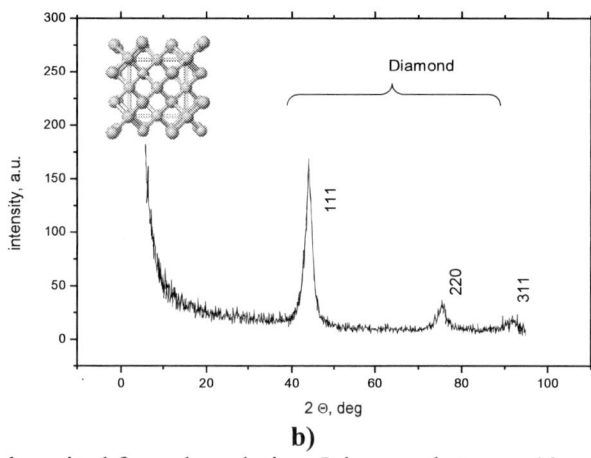

a)   b)

Figure 1. (a) TEM images of the nanodiamonds deposited from the solution. It is seen that near-10 nm nanodiamonds form the agglomerates inset shows the electron diffraction pattern (b) XRD pattern of the nanodiamonds powder. The XRD peak position corresponds to the bulk diamond structure (inset).

Size distribution of the used NDs has been performed from X-ray diffraction analysis. Typical X-ray diffraction pattern obtained on the NDs based powder is shown in figure 1b. Angle positions of the recorded peaks correspond to the cubic crystalline structure of diamond. According to Hall-Williamson's model, the NC size distribution was found to be centered at 8.3 ± 2.2 nm.

Since TEM shows the presence of agglomerates on the dried sample, size distribution of the nanoparticles' number was obtained using the dynamic light scattering (DLS) method (figure 2). Since it was shown in [6] 37 nm is the least detectable size for nanodiamonds, using DLS method, we can only conclude that we have no nanodiamonds agglomerates into the solution above 37 nm.

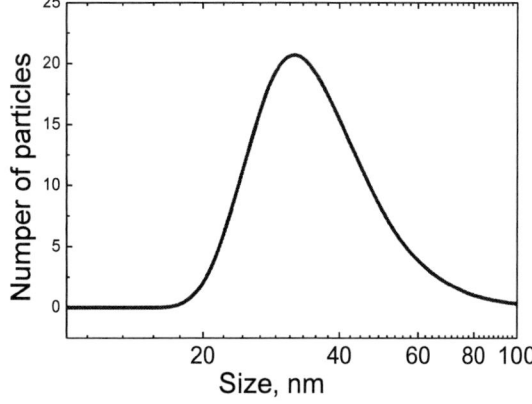

Figure 2. Size distribution of the nanoparticles' number obtained using the DLS of water suspensions of nanodiamonds.

Nanodiamonds are able to produce luminescence in green range under UV excitation. Figure 3 shows optical properties of the diamonds. Blue curve is an absorbance spectrum of the nanodiamonds, black one is a luminescence spectrum and the red is a PLE spectrum (intensity of PL at 490 nm under different excitation). The excited state lifetime is below 10 ns. The PL peak at 490 nm is attributed to the charge carriers recombination through the defects caused by nitrogen impurities, so the luminescence occurs predominantly due to recombination of charge carriers at the nitrogen-vacancy centers levels.

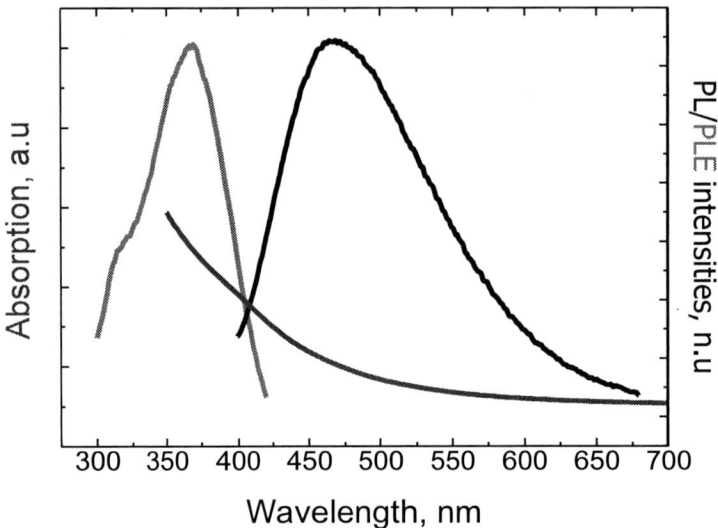

Figure 3. Photoluminescence and absorbance spectrum of the nanodiamonds (black and blue). PLE spectrum of the nanodiamonds (emission at 490 nm).

3. Toxicity estimation

In order to determine a lowest toxic concentration of the NDs they were added to the culture of 3T3-L1 living cells, and the in-time evolution of cell index (proportional to alive cell number) was recorded using non-destructive impedance-based method by means of an xCELLigence set-up (see protocols and methods section). The result of toxicity experiment can be seen on figure 4.

During first 48 h we can observe the 3T3-L1 proliferation, than at the pointed time, nanodiamonds with different concentrations were added to different cell wells. There is a dose-response of the cells' behavior after NPs addition. Highest concentration of NDs (2 g/L) decrease number of alive cells, intermediate ones slow down the proliferation rate and concentrations below 0.5 g/l have no effect on cells proliferation.

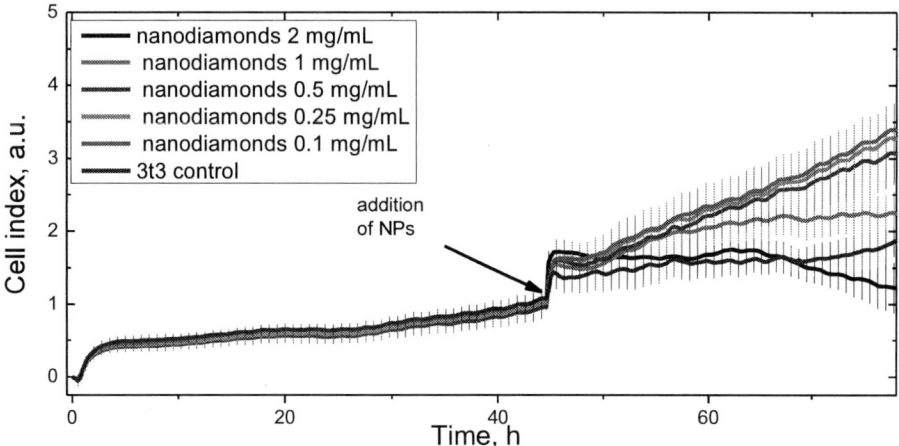

Figure 4. Cell index evolution curves for the 3T3-L1 cells. Arrow shows the time of the NDs addition.

4. One and two-photon excited imaging using NDs

Figure 6 shows bi-modal images of 3T3-L1 fibroblasts cells labelled with NDs. The labelling procedure is the 24h exposure of proliferating 3T3-L1 cells with 0.5 g/L of NDs. It is worth to notice the general healthy look of the cells, corroborating the non-toxic impact of the chosen NDs concentration. Figure 5a shows the Pl microscopy image of the fixed 3T3-L1 cells. Inset shows the PL of the control cells under the same conditions. We can observe the uniform distribution of the nanodiamonds inside the cells' cytoplasm, while the nuclei remain unlabelled.

In figure 5 b yellow colour is related to the two-photon excited luminescence. Since auto-fluorescence intensity of the non-labelled cells is too weak under these excitation/detection conditions (see inset in figure 5-b), the observed TPEL of the labelled cells can be attributed to the NDs accumulated inside the cellular cytoplasm and ensuring its quite bright staining.

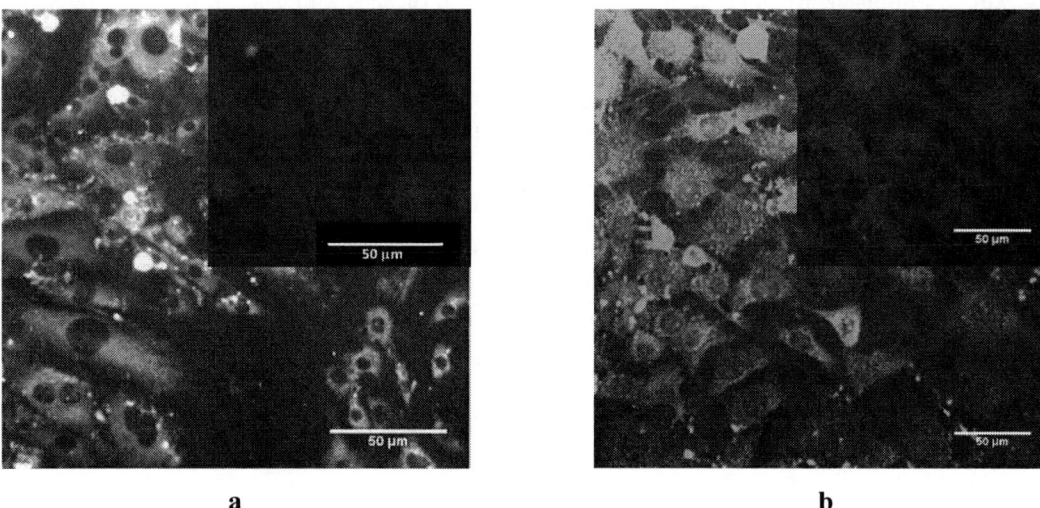

a   b

Figure 5. PL microscopy image of the 3T3-L1 cells labeled with 0.25 mg/mL of nanodiamonds for 24h.

5. Protocols and methods

TEM pictures were obtained by EM-002B (Topcon, Japan) high-resolution transmission electron microscope operating at 200 kV.

The x-ray diffraction experiments were performed using Cu k aloha x-ray source (0.15418 nm), SmartLab diffractometer at the nanoparticles dried powders.

For DLS size distribution measurements Zetasizer Nano Z zeta potential analyzer had been used. All the measurements were applied at 22^0 C in water solution.

3T3-L1 (fibroblasts) cells used for this work were grown in Dulbecco's modified Eagle's medium supplemented with 10% newborn calf serum 100 IU penicillin, 0.1mg streptomycin, and 0.25 mg/L amphotericin B at 37°C in a water saturated atmosphere with 5% CO_2 in air, in a Heraeus incubator. The cells were trypsinized and about 2500 cells (estimated by millpore Scepter™ 2.0 cell counter) were added to each well in a 96-well plate for cell proliferation rate measurements and onto the glass coverslip for cell imaging experiments. Then the cells were incubated for 48 h. NDs with concentrations 2, 1, 0.5, 0.25 and 0.1 mg/ml were added to the cell cultures which were additionally incubated for 24 h. After NDs were washed out from the extracellular environment with phosphate-buffered saline and the cells were washed with PBS twice. Than cells were fixed with 60% ethanol solution for the cell images acquisition.

Cell number measurements were performed using non-destructive impedance-based method (xCELLigence). The cells were grown on special plate with electrodes on its bottom. The system measures electrical impedance across interdigitated microelectrodes situated at the bottom of culture wells. The measurements are done by applying an alternative excitation signal (20 mV control voltage amplitude) at three different frequencies (10, 25 and 50 kHz) through the microelectrodes in the E-plates while monitoring the voltage drop across the electrodes where the quotient voltage/current yields the impedance. Software shows cell index as a result of processing impedance data. Cell index is proportional to the cell number, single cell surface area and adhesion factor. For a given cell line under basal conditions, cell number is the main factor affecting cell index. Each curve is the mean of cell index measured on 8 wells.

The luminescent microscopy images were obtained by means of Leica DMI 4000B microscope with the following filter combination: UV/violet excitation band: 354 – 424 nm and observation spectral range: >470 nm.

For optical nonlinear imaging we employed a Nikon A1R multiphoton upright microscope (NIE-Nikon) coupled with an Insight Deepsee tunable laser oscillator (Spectra-Physics, 120 fs, 80 MHz, 680 - 1300 nm). The nonlinear signals were epi-collected by a Nikon 25 water immersion objective (CFI75 APO, N.A.1.1) spectrally filtered by tailored pairs of dichroic mirrors and interference filters and acquired in parallel either by a normal photomultiplier (600 - 655 nm) or a GaAsP photomultiplier (385 - 492 nm).

6. Conclusion

Fluorescent nanodiamond is a new valuable tool for long-term labeling, imaging, tracking. In this work we have successfully show that 8 nm NDs produced by the LHDP method were used as biocompatible solid-state labels for bi-modal non-linear optical imaging *in vitro*.

Acknowledgments

This work was supported by Competitiveness Program of National Research Nuclear University MEPhI.

References

[1]   Rondin L, Dantelle G, Slablab A, Grosshans F, Treussart F, Bergonzo P, Perruchas S, Gacoin T, Chaigneau M, Chang H C, Jacques V and Roch J F 2010 *Phys. Rev. B* **82** 115449
[2]   Mochalin V N, Shenderova O, Ho D and Gogotsi Y 2012 *Nature Nanotech.* **7** 11-23
[3]   Mohan N, Chen C S, Hsieh H H, Wu Y C and Chang H C 2010 *Nano lett.* **10** 3692-99

[4]   Chang Y R *et al* 2008 *Nat. Nanotechnol.* **3** 284-8
[5]   Perevedentseva E, Peer D, Uvarov V, Zousman B and Levinson O 2015 *J. Nanosci. Nanotechnol.* **15** 1045–52
[6]   Colpin Y, Swan A, Zvyagin A V and Plakhotnik T 2006 *Opt. Lett.* **31** 625

Permittivity and permeability of semi-infinite metamaterial

O V Porvatkina, A A Tishchenko and M N Strikhanov

National Research Nuclear University MEPhI (Moscow Engineering Physics Institute), 31 Kashirskoye shosse, 115409, Moscow, Russia

E-mail: OVPorvatkina@mephi.ru, Tishchenko@mephi.ru

Abstract. In our work we investigate dielectric and magnetic properties of semi-infinite metamaterial consisting of particles of different possible nature: atoms, molecules, nanoparticles, etc. It is important that these particles would have magnetic properties. Polarization of a near-surface layer is known to differ from its bulk value for non-magnetic materials; for magnetic materials, including metamaterials, the situation should be similar, which is the subject of our research. We obtain analogues of the Clausius-Mossotti relation both for permittivity and permeability taking into account the local field effects in the long-wave approximation for semi-infinite metamaterial. These relations describe the connection between macroscopic characteristics of the semi-infinite metamaterial (permittivity and permeability) and characteristics of constituent particles (dielectric polarizability and magnetic polarizability), which is a bright example of multi-scale approach - method very popular today in physical and computer simulating.

1. Introduction

In recent years engineered materials composed of designed inclusions have drawn significant scientific interest, underscoring the remarkable potential of them to broaden the range of possible wave phenomena not usually observed in nature but being accessible via laboratory experiment [1-3]. These artificially structured composites, known as metamaterials, demonstrate unusual electromagnetic and optical properties, e.g., negative refractive index and subwavelength focusing [4-6]. The properties of metamaterials are derived both from the intrinsic properties of their constituent particles, as well as from the geometrical arrangement of those particles. A striking instance of this fact is a metamaterial constructed from insulating magnetodielectric spherical particles embedded in a background dielectric material [7,8].

Due to their properties metamaterials are in the centre of modern investigations. They find use in various applications. For instance, metamaterial coatings have been employed to enhance the radiation and matching properties of electrically small electric and magnetic dipole antennas [9] and for engineering sensors with specified sensitivity [10]. Metamaterials are very interesting for researches because of their unique electromagnetic properties, which are not realizable in natural materials [11,12]. Most of interesting theoretical problems are connected with the boundary of semi-infinite metamaterial. On the other hand, the local field effects have a strong influence on electromagnetic processes in medium and therefore can be of great interest for different applications in optics, including nonlinear optics and optics of metamaterials [13-15]. In this work we consider a semi-infinite metamaterial starting from the first principles and find its dielectric and magnetic properties on the base of microscopic structure and properties of single particles, which is possible only with help of the local field method.

Content from this work may be used under the terms of the Creative Commons Attribution 3.0 licence. Any further distribution of this work must maintain attribution to the author(s) and the title of the work, journal citation and DOI.

Published under licence by IOP Publishing Ltd

2. Dielectric and magnetic properties of semi-infinite metamaterial

We consider metamaterial, which occupies the semi-infinite space $z > 0$. This material is composed of N anisotropic particles. The particles have both dielectric and magnetic properties:

$$\alpha_{ij}^{e}(\omega) = \alpha_{\perp}^{e}(\omega)\left(\delta_{ij} - e_i e_j\right) + \alpha_{l}^{e}(\omega) e_i e_j, \tag{1}$$

$$\alpha_{ij}^{m}(\omega) = \alpha_{\perp}^{m}(\omega)\left(\delta_{ij} - e_i e_j\right) + \alpha_{l}^{m}(\omega) e_i e_j. \tag{2}$$

The electromagnetic field $(\mathbf{E}_0, \mathbf{B}_0)$ is acting on semi-infinite metamaterial. The Fourier-transform for density of microcurrents in this medium has the form:

$$\left\langle j_i^{mic}(\mathbf{r}',\omega)\right\rangle = -i\omega\alpha_{ij}^{e}(\omega)\left\langle \sum_n E_j^{mic}(\mathbf{R}_n,\omega)\delta(\mathbf{r}'-\mathbf{R}_n)\right\rangle$$
$$+ c\alpha_{ij}^{m}(\omega)\left\langle \sum_n \left(rot\mathbf{H}^{mic}(\mathbf{R}_n,\omega)\right)_j \delta(\mathbf{r}'-\mathbf{R}_n)\right\rangle, \tag{3}$$

We do our calculations taking into account the so called local field effects, meaning that the local field is the field acting on the particle of the medium averaged over the positions of other particles [16]. These effects have a strong influence on the optical nonlinear phenomena, in particular, on nonlinear optics of planar metamaterials [17]. We use relation between macroscopic (electric and magnetic) fields and local (electric and magnetic) fields for 3D metamaterials obtained in our work [18]. So, we obtain expressions relating macroscopic and local fields for semi-infinite metamaterials:

$$E_i(\mathbf{p},\omega) = G_{ij}(\mathbf{p},\omega) E_j^{loc}(\mathbf{p},\omega), \tag{4}$$

$$B_i(\mathbf{p},\omega) = F_{ij}(\mathbf{p},\omega) B_j^{loc}(\mathbf{p},\omega), \tag{5}$$

where

$$G_{ij}(\mathbf{p},\omega) = \delta_{ij} + \frac{n}{2\pi^2}\int d^3 p \frac{k^2\delta_{ik} - p_i p_k}{p^2 - k^2}\int_{z'>0} d^3 R' \exp(i\mathbf{p}\mathbf{R}') f(\mathbf{R}')\alpha_{kj}^{e}(\omega), \tag{6}$$

$$F_{ij}(\mathbf{p},\omega) = \delta_{ij} + \frac{n}{2\pi^2}\int d^3 p \frac{p^2\delta_{ik} - p_i p_k}{p^2 - k^2}\int_{z'>0} d^3 R' \exp(i\mathbf{p}\mathbf{R}') f(\mathbf{R}')\alpha_{kj}^{m}(\omega). \tag{7}$$

Let us introduce the notations

$$Q_{ij}(Z) = \frac{n}{2\pi^2}\int d^3 p \frac{k^2\delta_{ik} - p_i p_k}{p^2 - k^2}\int_{z'>0} d^3 R'' \exp(i\mathbf{p}\mathbf{R}') f(\mathbf{R}')\alpha_{kj}^{e}(\omega), \tag{8}$$

$$S_{ij}(Z) = \frac{n}{2\pi^2}\int d^3 p \frac{p^2\delta_{ik} - p_i p_k}{p^2 - k^2}\int_{z'>0} d^3 R' \exp(i\mathbf{p}\mathbf{R}') f(\mathbf{R}')\alpha_{kj}^{m}(\omega). \tag{9}$$

For infinite material these tensors become simpler

$$Q_{ij}(Z) = -\frac{1}{3}\delta_{ij}, \quad S_{ij}(Z) = \frac{2}{3}\delta_{ij}. \tag{10}$$

Using these values, one can obtain equations (8), (9) in the form

$$Q_{ij}(Z) = -\frac{1}{3}\delta_{ij} - W_{ij}(z), \quad S_{ij}(Z) = \frac{2}{3}\delta_{ij} - V_{ij}(z), \tag{11}$$

where

$$W_{ij}(Z) = \frac{n}{2\pi^2}\int d^3 p \frac{k^2\delta_{ik} - p_i p_k}{p^2 - k^2}\int_{z'<0} d^3 R' \exp(i\mathbf{p}\mathbf{R}') f(\mathbf{R}')\alpha_{kj}^{e}(\omega), \tag{12}$$

$$V_{ij}(Z) = \frac{n}{2\pi^2} \int d^3p \frac{p^2 \delta_{ik} - p_i p_k}{p^2 - k^2} \int\limits_{z'<0} d^3R' \exp(i\mathbf{pR'}) f(\mathbf{R'}) \alpha_{kj}^m(\omega). \tag{13}$$

Then equations (6),(7) can be rewritten as

$$G_{ij}(\mathbf{p},\omega) = \frac{2}{3}\delta_{ij} - W_{ij}(z), \quad F_{ij}(\mathbf{p},\omega) = \frac{5}{3}\delta_{ij} - V_{ij}(z). \tag{14}$$

In equations (14) $f(\mathbf{R'})$ is the distribution function, which depends on properties of the media and can be found experimentally [19].

It is well known that electric (magnetic) induction is a vector quantity that equals to the sum of the vector of electric (magnetic) field and the polarization (magnetization) of the medium:

$$D_i(\mathbf{p},\omega) = E_i(\mathbf{p},\omega) + 4\pi P_i(\mathbf{p},\omega), \tag{15}$$

$$H_i(\mathbf{p},\omega) = B_i(\mathbf{p},\omega) - 4\pi M_i(\mathbf{p},\omega). \tag{16}$$

Considering the fact that polarizability and magnetization can be expressed in terms of both macroscopic and local fields, we can write

$$\left(\varepsilon_{ij}(\mathbf{p},\omega) - \delta_{ij}\right) E_j(\mathbf{p},\omega) = 4\pi n\alpha_{ij}^e(\omega) E_j^{loc}(\mathbf{p},\omega), \tag{17}$$

$$\left(\mu_{ij}^{-1}(\mathbf{p},\omega) - \delta_{ij}\right) B_j(\mathbf{p},\omega) = -4\pi n\alpha_{ij}^m(\omega) B_j^{loc}(\mathbf{p},\omega). \tag{18}$$

The equations for permittivity and permeability are obtained by substituting local fields from equations (4), (5) in equations (17), (18) which gives:

$$\varepsilon_{ik}(\mathbf{p},\omega) = \delta_{ik} + 4\pi n\alpha_{ij}^e(\omega) G_{jk}^{-1}(\mathbf{p},\omega), \tag{19}$$

$$\mu_{ik}^{-1}(\mathbf{p},\omega) = \delta_{ik} - 4\pi n\alpha_{ij}^m(\omega) F_{jk}^{-1}(\mathbf{p},\omega). \tag{20}$$

Finally, using equation(1) we get:

$$\varepsilon_{ik}(\mathbf{p},\omega) = \delta_{ik} + 4\pi n\alpha_{\perp}^e(\omega) G_{ik}^{-1}(\mathbf{p},\omega) + 4\pi n\left(\alpha_{l}^e(\omega) - \alpha_{\perp}^e(\omega)\right) e_i G_{zk}^{-1}(\mathbf{p},\omega), \tag{21}$$

$$\mu_{ik}^{-1}(\mathbf{p},\omega) = \delta_{ik} - 4\pi n\alpha_{\perp}^m(\omega) F_{ik}^{-1}(\mathbf{p},\omega) - 4\pi n\left(\alpha_{l}^m(\omega) - \alpha_{\perp}^m(\omega)\right) e_i F_{zk}^{-1}(\mathbf{p},\omega). \tag{22}$$

These equations are analogue of the Clausius-Mossotti relations for semi-infinite metamaterial.

Equations (21), (22) describe dielectric and magnetic properties of semi-infinite metamaterial in the near-surface region. In these equations the tensors depend on the radial distribution function $f(\mathbf{R'})$, which can be measured using x-ray diffraction.

As long as we neglect the effects of the polarization of a near-surface layer, it is easy to get the expressions which turn into results obtained in the work [20] for anisotropic non-magnetic media ($\mu = 1$).

In the long-wave limit

$$n^{-1/3} << q^{-1} << c/\omega \tag{23}$$

for metamaterials consisting of spherically symmetric particles tensors $W_{ij}(Z), V_{ij}(Z)$ have only one preferred direction (Z axis). These tensors are axially symmetric with respect to the Z axis. For this reason components $W_{zz}(Z)$ and $V_{zz}(Z)$ differs from components $W_{xx}(Z) = W_{yy}(Z)$ and $V_{xx}(Z) = V_{yy}(Z)$.

In case of infinite material with magnetic properties from equations (19), (20) we get:

$$\varepsilon(\omega) = \frac{1 + (8\pi/3)n\alpha_e(\omega)}{1 - (4\pi/3)n\alpha_e(\omega)}, \tag{24}$$

$$\mu(\omega) = \frac{1 + (8\pi/3)n\alpha_m(\omega)}{1 - (4\pi/3)n\alpha_m(\omega)}. \tag{25}$$

Equations (24), (25) coincide with both results obtained in our previous work [18] and with results from [21] obtained with the help of the other method.

So, using the Clausius-Mossotti relation, one can transform our results to the following form:

$$\varepsilon_{xx}\left(z,\omega\right)=\varepsilon_{yy}\left(z,\omega\right)=\frac{\varepsilon\left(\omega\right)+\left(\varepsilon\left(\omega\right)-1\right)W_{xx}\left(z\right)}{1+\left(\varepsilon\left(\omega\right)-1\right)W_{xx}\left(z\right)},\tag{26}$$

$$\varepsilon_{zz}\left(z,\omega\right)=\frac{\varepsilon\left(\omega\right)+\left(\varepsilon\left(\omega\right)-1\right)W_{zz}\left(z\right)}{1+\left(\varepsilon\left(\omega\right)-1\right)W_{zz}\left(z\right)},\tag{27}$$

$$\mu_{zz}\left(z,\omega\right)=\frac{\mu\left(\omega\right)+\left(\mu\left(\omega\right)-1\right)V_{zz}\left(z\right)}{1+\left(\mu\left(\omega\right)-1\right)V_{zz}\left(z\right)},\tag{28}$$

$$\mu_{xx}\left(z,\omega\right)=\mu_{yy}\left(z,\omega\right)=\frac{\mu\left(\omega\right)+\left(\mu\left(\omega\right)-1\right)V_{xx}\left(z\right)}{1+\left(\mu\left(\omega\right)-1\right)V_{xx}\left(z\right)},\tag{29}$$

where

$$W_{xx}\left(z\right)=\frac{n\alpha^{e}\left(\omega\right)}{2\pi^{2}}\int d^{3}p\frac{k^{2}-p_{x}^{2}}{p^{2}-k^{2}}\int\limits_{z'<0}d^{3}R'\exp\left(i\mathbf{p}\mathbf{R}'\right)f\left(\mathbf{R}'\right),\tag{30}$$

$$W_{zz}\left(z\right)=\frac{n\alpha^{e}\left(\omega\right)}{2\pi^{2}}\int d^{3}p\frac{k^{2}-p_{z}^{2}}{p^{2}-k^{2}}\int\limits_{z'<0}d^{3}R'\exp\left(i\mathbf{p}\mathbf{R}'\right)f\left(\mathbf{R}'\right),\tag{31}$$

$$V_{xx}\left(Z\right)=\frac{n\alpha^{m}\left(\omega\right)}{2\pi^{2}}\int d^{3}p\frac{p^{2}-p_{x}^{2}}{p^{2}-k^{2}}\int\limits_{z'<0}d^{3}R'\exp\left(i\mathbf{p}\mathbf{R}'\right)f\left(\mathbf{R}'\right),\tag{32}$$

$$V_{zz}\left(Z\right)=\frac{n\alpha^{m}\left(\omega\right)}{2\pi^{2}}\int d^{3}p\frac{p^{2}-p_{z}^{2}}{p^{2}-k^{2}}\int\limits_{z'<0}d^{3}R'\exp\left(i\mathbf{p}\mathbf{R}'\right)f\left(\mathbf{R}'\right).\tag{33}$$

Equations (26)-(29) describe dielectric properties of near-surface layer macroscopically, using permittivity tensor depending on the coordinate Z. The result for the permittivity coincides with those from the work [22].

3. Discussion

In this paper dielectric and magnetic properties of semi-infinite metamaterials were investigated with taking into account the local field effects. We have obtained equations for permittivity tensor (21) and permeability tensor (22) in the near-surface region. These tensors allow analyzing dielectric and magnetic properties of semi-infinite metamaterials and considering qualitatively new phenomena in metamaterials. Thus, using these equations for permittivity and permeability one can construct semi-infinite metamaterials based on the properties of their constituent particles of different shapes and nature. For example, in our recent work [23] we obtained an analogue of the Clausius-Mossotti relations for 3D metamaterials based on colloidal quantum dots.

We have obtained the permittivity and permeability tensors (26)-(29) in long-wave limit in case of spherically symmetric particles as well.

The results obtained are important for understanding the influence of polarization and magnetization of the near-surface layer of metamaterial on the properties of electromagnetic surface waves, reflection and refraction and other optical phenomena. Also, the expressions for permittivity and permeability can serve as the theoretical foundation for engineering and designing general metamaterial-based devices.

Acknowledgments

This work was supported by the Ministry of Education and Science of the Russian Federation, the project 3.1110.2014/K and the Competitiveness Programm of National Research Nuclear University MEPhI.

References

[1] Houck A A, Brock J B and Chuang I L 2003 *Phys. Rev. Lett.* **90** 137401

[2] Luo C, Johnson S G, Joannopoulos J D and Pendry J B 2002 *Phys. Rev.* B **65** 201104

[3] Wheeler M S, Aitchison J S and Mojahedi M 2006 *Phys. Rev.* B **73** 045105

[4] Pendry J B 2000 *Phys. Rev. Lett.* **85** 3966-9

[5] Smith D R, Pendry J B and Wiltshire M C K 2004 *Science* **305** 788-92

[6] Smith D R, Padilla W J, Vier D C, Nemat-Nasser S C and Schultz S 2000 Phys. Rev. Lett. **84** 4184-7

[7] Enkrich C, Wegener M, Linden S, Burger S, Zschiedrich L, Schmidt F, Zhou J F, Koschny Th and Soukoulis C M 2005 *Phys. Rev. Lett.* **95** 203901

[8] Parimi P V, Lu W T, Vodo P and Sridhar S 2003 *Nature* **426** 404

[9] Lapine M and Tretyakov S 2007 *IET Microw. Antennas Propag.,* **1** 3–11

[10] Gangwar K, Paras and Gangwar R P S 2014 *AEEE* **4** 97-106

[11] Bankov S E 2008 *IJECT* **53** 15–25

[12] Mochan W L and Barrera R G 1985 *Phys. Lev. Lett.* **55** 1192-5

[13] Kästel J and Fleischhauer M 2007 *Phys. Rev.* A **76** 062509

[14] Gardner D F, Evans J S and Smalyukh I I 2011 *Molecular Crystals and Liquid Crystals* **545** 1221-45

[15] Linden S, Enkrich C, Wegener M, Zhou J, Koschny T and Soukoulis C 2004 *Science* **306** 1351-3

[16] Gorkunov M V and Ryazanov M I 1997 *JETP* **85** 97-103

[17] Porvatkina O V, Tishchenko A A and Strikhanov M N 2015 Progress in Electromagnetics Research Symposium (PIERS) Proc. (Prague) vol 2015 (Cambridge, MA:The Electromagnetics Academy) pp1689-92

[18] Porvatkina O V, Tishchenko A A, Ryazanov M I and Strikhanov M N 2014 *J. Phys.: Conf. Ser.* **541** 012024

[19] Ryazanov M I 1984 *Electrodynamics of Condensed Matter* (Moscow: Nauka) (in Russian)

[20] Anokhin M N, Tishchenko A A, Ryazanov M I and Strikhanov M N 2014 *J. Phys.: Conf. Ser.* **541** 012023

[21] Kästel J and Fleischhauer M 2007 *Physical Review* A **76** 062509

[22] Ryazanov M I 1996 *JETP* **83** 529-32

[23] Porvatkina O V, Tishchenko A A and Strikhanov M N 2015 *J. Phys.: Conf. Ser.* **643** 012074

CORSCS2015　　　　　　　　　　　　　　　　　　　　　　　IOP Publishing
Journal of Physics: Conference Series **740** (2016) 012012　　doi:10.1088/1742-6596/740/1/012012

Generalized Clausius-Mossotti relation for semi-infinite artificial periodic structure

M N Anokhin, A A Tishchenko and M N Strikhanov

National Research Nuclear University MEPhI (Moscow Engineering Physics Institute), 31 Kashirskoye shosse, 115409, Moscow, Russia

E-mail: MNAnokhin@mephi.ru, Tishchenko@mephi.ru

Abstract. We obtain from the first principles a generalized Clausius-Mossotti relation describing the dielectric permittivity of a semi-infinite artificial periodic structure. The obtained expressions include the spatial dispersion and permit defining resonant conditions for propagating waves.

1. Introduction

Periodicity changes the dielectric properties and, consequently, determines the propagation of electromagnetic waves in periodic structures of various types [1-3]. Periodic structures are widely used in different applications, e.g., in new and perspective class of materials – photonic crystals and metamaterials [4], for producing high-performance filters, in resonators, signal dividers, microwave electronics, etc.

Photonic crystal is a periodic structure which allows controlling light by opening a bandgap within a range of forbidden frequencies. Theoretical calculations of a light propagation in photonic crystals are based on the general theory for periodic structures [5].

In this work we develop the so called local field theory for the case of semi-infinite artificial periodic structure. In our recent paper [6] we considered an infinite structure and now demonstrate that existence of the surface leads to the additional anisotropy and, thus, changes the tensor structure of the dielectric permittivity. The method we use is based on the direct solving Maxwell's equations, and it is known that in case of amorphous medium the natural changing of the dielectric properties near the surface occurs [7, 8].

2. Dielectric properties of semi-infinite artificial periodic structure

We consider the semi-infinite periodic structure occupying half-space $z > 0$ composed of N anisotropic particles with the same polarizability $\alpha_{ij}(\omega)$:

$$\alpha_{ij}(\omega) = \alpha_\perp(\omega)(\delta_{ij} - e_i e_j) + \alpha_\parallel(\omega) e_i e_j. \tag{1}$$

Let the external field $\mathbf{E}^0$ act on this structure. One can write a solution of the Fourier transform of Maxwell's equations in a medium in the dipole approximation. The microscopic field acting on the a-th particle can be written as [6, 9]:

(cc) Content from this work may be used under the terms of the Creative Commons Attribution 3.0 licence. Any further distribution of this work must maintain attribution to the author(s) and the title of the work, journal citation and DOI.

Published under licence by IOP Publishing Ltd

$$E_i^{mic}\left(\mathbf{R}_a,\omega\right)=E_i^0\left(\mathbf{R}_a,\omega\right)+$$
$$+\frac{1}{2\pi^2}\int d^3 l S_{ij}\left(\mathbf{l},\omega\right)\alpha_{jk}\left(\omega\right)\sum_b E_k^{mic}\left(\mathbf{R}_b,\omega\right)\exp\left\{-i\mathbf{l}\left(\mathbf{R}_b-\mathbf{R}_a\right)\right\}, \tag{2}$$

where

$$S_{ij}\left(\mathbf{l},\omega\right)=\frac{k^2\delta_{ij}-l_i l_j}{l^2-k^2-i0}, \tag{3}$$
$$k=\omega/c, \tag{4}$$

where index b corresponds to all the other particles of this structure with the exception of a-th particle. Equation (2) can be solved only approximately because of $N\gg1$.

The addend in equation (2) is formed by the sum of the fields of the rest particles of matter. The contribution of any particle depends on its position relative to the a-th particle. It determines the dependence of the effective field on the mutual arrangement of the particles, i.e. in fact the structure of matter.

Let $w\left(\mathbf{R}_{ba}\right)$ be the probability density of finding the b-th particle at the distance $\mathbf{R}_{ba}=\mathbf{R}_b-\mathbf{R}_a$ $\left(k=\overline{1,N-1}\right)$ from the a-th one:

$$w\left(\mathbf{R}_{ba}\right)=\frac{1}{N}\sum_{m=1}^{N}\delta\left(\mathbf{R}_{ba}-\mathbf{R}_m\right). \tag{5}$$

Let us replace $\mathbf{E}^{mic}$ in the first approximation in equation (2) with its averaged over other particles value called the local field $\mathbf{E}^{loc}$:

$$E_i^{loc}\left(\mathbf{R},\omega\right)=E_i^0\left(\mathbf{R},\omega\right)+\frac{1}{2\pi^2}\sum_{m=1}^{N}S_{ij}\left(-\mathbf{R}_m,\omega\right)\alpha_{jk}\left(\omega\right)E_k^{loc}\left(\mathbf{R}+\mathbf{R}_m,\omega\right), \tag{6}$$

where

$$S_{ij}\left(-\mathbf{R},\omega\right)=\int d^3 l S_{ij}\left(\mathbf{l},\omega\right)\exp\left\{-i\mathbf{l}\mathbf{R}\right\}. \tag{7}$$

We take into account that all the particles are located only in the region $z>0$.

One can find the macroscopic field by averaging equation (2) over the coordinates of all the particles:

$$E_i\left(\mathbf{R},\omega\right)=E_i^0\left(\mathbf{R},\omega\right)+$$
$$+\frac{1}{2\pi^2}\int d^3 l S_{ij}\left(\mathbf{l},\omega\right)\alpha_{jk}\left(\omega\right)\left\langle\sum_b E_k^{loc}\left(\mathbf{R}_b,\omega\right)\exp\left\{-i\mathbf{l}\left(\mathbf{R}_b-\mathbf{R}\right)\right\}\right\rangle, \tag{8}$$

where

$$\left\langle\sum_b E_k^{loc}\left(\mathbf{R}_b,\omega\right)\exp\left\{-i\mathbf{l}\left(\mathbf{R}_b-\mathbf{R}\right)\right\}\right\rangle=\frac{N}{V}\int\limits_{z''>0}d^3\mathbf{R}''E_k^{loc}\left(\mathbf{R}'',\omega\right)\exp\left\{-i\mathbf{l}\left(\mathbf{R}''-\mathbf{R}\right)\right\}. \tag{9}$$

Then, the macroscopic field can be obtained in form

$$E_i\left(\mathbf{R},\omega\right)=E_i^0\left(\mathbf{R},\omega\right)+\frac{1}{2\pi^2}n\int\limits_{z'+z>0}d^3\mathbf{R}'S_{ij}\left(-\mathbf{R}',\omega\right)\alpha_{jk}\left(\omega\right)E_k^{loc}\left(\mathbf{R}'+\mathbf{R},\omega\right). \tag{10}$$

The macroscopic field is expressed through the local field using equations (6) and (10):

$$E_i\left(\mathbf{R},\omega\right)=E_i^{loc}\left(\mathbf{R},\omega\right)+\frac{1}{2\pi^2}n\int\limits_{z'+z>0}d^3\mathbf{R}'S_{ij}\left(-\mathbf{R}',\omega\right)\alpha_{jk}\left(\omega\right)E_k^{loc}\left(\mathbf{R}'+\mathbf{R},\omega\right)-$$
$$-\frac{1}{2\pi^2}\sum_{m=1}^{N}S_{ij}\left(-\mathbf{R}_m,\omega\right)\alpha_{jk}\left(\omega\right)E_k^{loc}\left(\mathbf{R}+\mathbf{R}_m,\omega\right). \tag{11}$$

Equation (11) can be written in variables $\left(\mathbf{q},\omega,z\right)$:

$$E_i\left(\mathbf{q},z,\omega\right) = E_i^{loc}\left(\mathbf{q},z,\omega\right) +$$

$$+\frac{1}{2\pi^2}n\int\limits_{Z'+Z>0} d^3\mathbf{R}'\exp\{i\mathbf{qR}'\}S_{ij}\left(-\mathbf{R}',\omega\right)\alpha_{jk}\left(\omega\right)E_k^{loc}\left(\mathbf{q},z+Z',\omega\right) - \tag{12}$$

$$-\frac{1}{2\pi^2}\sum_{m=1}^{N}S_{ij}\left(-\mathbf{R}_m,\omega\right)\alpha_{jk}\left(\omega\right)E_k^{loc}\left(\mathbf{q},z+Z_m,\omega\right)\exp\{i\mathbf{qR}_m\}.$$

A relation between the Fourier transforms of the local and macroscopic fields [10]:

$$t_{ik}\left(\mathbf{q},z,\omega\right)E_k^{loc}\left(\mathbf{q},z,\omega\right) = E_i\left(\mathbf{q},z,\omega\right), \tag{13}$$

where

$$t_{ik}\left(\mathbf{q},z,\omega\right) = \delta_{ik} + \frac{1}{2\pi^2}n\int\limits_{Z'+Z>0} d^3\mathbf{R}'\exp\{i\mathbf{qR}'\}S_{ij}\left(-\mathbf{R}',\omega\right)\alpha_{jk}\left(\omega\right) -$$

$$-\frac{1}{2\pi^2}\sum_{m=1}^{N}\exp\{i\mathbf{qR}_m\}S_{ij}\left(-\mathbf{R}_m,\omega\right)\alpha_{jk}\left(\omega\right). \tag{14}$$

Let us make some auxiliary manipulations:

$$S_{ij}\left(-\mathbf{R},\omega\right) = \int d^3l\, S_{ij}\left(\mathbf{l},\omega\right)\exp\{-i\mathbf{lR}\} = a(R)\delta_{ij} + b(R)\frac{R_iR_j}{R^2}. \tag{15}$$

Thus, tensor $t_{ik}\left(\mathbf{q},z,\omega\right)$ has the form:

$$t_{ik}\left(\mathbf{q},z,\omega\right) = \delta_{ik} + \frac{1}{2\pi^2}n\int\limits_{Z'+Z>0} d^3\mathbf{R}'\exp\{i\mathbf{qR}'\}\left\{a(R')\delta_{ij} + b(R')\frac{R_i'R_j'}{R'^2}\right\}\alpha_{jk}\left(\omega\right) -$$

$$-\frac{1}{2\pi^2}\sum_{m=1}^{N}\exp\{i\mathbf{qR}_m\}\left\{a(R_m)\delta_{ij} + b(R_m)\frac{R_i^mR_j^m}{R_m^2}\right\}\alpha_{jk}\left(\omega\right), \tag{16}$$

where

$$a(R) = -2\frac{\pi^2k}{R^2}\sin(kR) - 2\frac{\pi^2}{R^3}\cos(kR);$$

$$b(R) = 4\frac{\pi^2k^2}{R}\cos(kR) + 6\frac{\pi^2k}{R^2}\sin(kR) + 6\frac{\pi^2}{R^3}\cos(kR). \tag{17}$$

As a result, we find expression for the dielectric permittivity:

$$\varepsilon_{ik}\left(\mathbf{q},z.\omega\right) = \delta_{ik} + 4\pi n\alpha_{ij}\left(\omega\right)t_{jk}^{-1}\left(\mathbf{q},z,\omega\right), \tag{18}$$

where $z>0$.

Let us neglect the anisotropy of individual particles, i.e.:

$$\alpha\left(\omega\right) \equiv \alpha_\perp\left(\omega\right) \equiv \alpha_\parallel\left(\omega\right). \tag{19}$$

We can find a tensor $t_{ik}^{-1}\left(\mathbf{q},z,\omega\right)$ by presenting tensor $t_{ik}\left(\mathbf{q},z,\omega\right)$ in the form:

$$t_{ik}\left(\mathbf{q},z,\omega\right) = c_1\left(\mathbf{q},z,\omega\right)\delta_{ik} + c_2\left(\mathbf{q},z,\omega\right)\frac{q_iq_k}{q^2} +$$

$$+c_3\left(\mathbf{q},z,\omega\right)e_ie_k + c_4\left(\mathbf{q},z,\omega\right)e_iq_k + c_5\left(\mathbf{q},z,\omega\right)q_ie_k. \tag{20}$$

From the condition

$$t_{ik}\left(\mathbf{q},z,\omega\right) = t_{ki}\left(\mathbf{q},z,\omega\right), \tag{21}$$

one can see that

$$c_4\left(\mathbf{q},z,\omega\right) = c_5\left(\mathbf{q},z,\omega\right). \tag{22}$$

Therefore, equation (20) goes to

$$t_{ik}(\mathbf{q},z,\omega) = c_1(\mathbf{q},z,\omega)\delta_{ik} + c_2(\mathbf{q},z,\omega)\frac{q_i q_k}{q^2} +$$
$$+ c_3(\mathbf{q},z,\omega)e_i e_k + c_4(\mathbf{q},z,\omega)(e_i q_k + q_i e_k). \tag{23}$$

Comparing equations (16) and (23) we find the coefficient $c_1(\mathbf{q},z,\omega)$:

$$c_1(\mathbf{q},z,\omega) = 1 + a_1(\mathbf{q},z,\omega) - a_2(\mathbf{q},z,\omega), \tag{24}$$

where

$$a_1(\mathbf{q},z,\omega) = \frac{1}{2\pi^2}n\alpha(\omega)\int\limits_{Z'+Z>0} d^3\mathbf{R}'\exp\{i\mathbf{q}\mathbf{R}'\}a(R');$$

$$a_2(\mathbf{q},z,\omega) = \frac{1}{2\pi^2}\alpha(\omega)\sum_{m=1}^{N}\exp\{i\mathbf{q}\mathbf{R}_m\}a(R_m). \tag{25}$$

The other coefficients can be obtained by multiplying equation (23) by δ_{ki}, $e_k e_i$, $e_k q_i$:

$$c_2(\mathbf{q},z,\omega) = b_1(\mathbf{q},z,\omega) - b_2(\mathbf{q},z,\omega);$$
$$c_3(\mathbf{q},z,\omega) = b_3(\mathbf{q},z,\omega) - b_4(\mathbf{q},z,\omega); \tag{26}$$
$$c_4(\mathbf{q},z,\omega) = b_5(\mathbf{q},z,\omega) - b_6(\mathbf{q},z,\omega),$$

where

$$b_1(\mathbf{q},z,\omega) = \frac{1}{2\pi^2}n\alpha(\omega)\int\limits_{Z'+Z>0} d^3\mathbf{R}'\exp\{i\mathbf{q}\mathbf{R}'\}b(R')\left[1 - \frac{Z'^2}{R'^2}\right];$$

$$b_2(\mathbf{q},z,\omega) = \frac{1}{2\pi^2}\alpha(\omega)\sum_{m=1}^{N}\exp\{i\mathbf{q}\mathbf{R}_m\}b(R_m)\left[1 - \frac{Z_m'^2}{R_m^2}\right];$$

$$b_3(\mathbf{q},z,\omega) = \frac{1}{2\pi^2}n\alpha(\omega)\int\limits_{Z'+Z>0} d^3\mathbf{R}'\exp\{i\mathbf{q}\mathbf{R}'\}b(R')\frac{Z'^2}{R'^2};$$

$$b_4(\mathbf{q},z,\omega) = \frac{1}{2\pi^2}\alpha(\omega)\sum_{m=1}^{N}\exp\{i\mathbf{q}\mathbf{R}_m\}b(R_m)\frac{Z_m'^2}{R_m^2}; \tag{27}$$

$$b_5(\mathbf{q},z,\omega) = \frac{1}{2\pi^2}n\alpha(\omega)\int\limits_{Z'+Z>0} d^3\mathbf{R}'\exp\{i\mathbf{q}\mathbf{R}'\}b(R')\frac{(\mathbf{q}\mathbf{R})Z_k'}{q^2 R'^2};$$

$$b_6(\mathbf{q},z,\omega) = \frac{1}{2\pi^2}\alpha(\omega)\sum_{m=1}^{N}\exp\{i\mathbf{q}\mathbf{R}_m\}b(R_m)\frac{(\mathbf{q}\mathbf{R})Z_m}{q^2 R_m^2}.$$

Tensor $t_{ik}^{-1}(\mathbf{q},z,\omega)$ has the same structure as $t_{ik}(\mathbf{q},z,\omega)$ in equation (23):

$$t_{ik}^{-1}(\mathbf{q},z,\omega) = d_1(\mathbf{q},z,\omega)\delta_{ik} + d_2(\mathbf{q},z,\omega)\frac{q_i q_k}{q^2} +$$
$$+ d_3(\mathbf{q},z,\omega)e_i e_k + d_4(\mathbf{q},z,\omega)(e_i q_k + q_i e_k). \tag{28}$$

For finding the coefficients $d_{1,2,3,4}(\mathbf{q},z,\omega)$ in equation (28) it is necessary to carry out some additional calculations:

$$t_{ki}^{-1}(\mathbf{q},z,\omega)t_{ij}(\mathbf{q},z,\omega) = \delta_{kj}. \tag{29}$$

To make the calculations easier, let us put

$$c_\alpha(\mathbf{q},z,\omega) \equiv c_\alpha;$$
$$d_\alpha(\mathbf{q},z,\omega) \equiv d_\alpha, \tag{30}$$

where $\alpha = 1,2,3,4$.

The tensor coefficients in equation (29) are grouped:

$$\delta_{kj} = d_1 c_1 \delta_{kj} + \left[d_1 c_2 + d_2 c_1 + d_2 c_2 + d_4 c_4 q^2 \right] \frac{q_k q_j}{q^2} + \left[d_1 c_3 + d_3 c_1 + d_3 c_3 + d_4 c_4 q^2 \right] e_k e_j +$$

$$+ \left[d_1 c_4 + d_3 c_4 + d_4 c_1 + d_4 c_2 \right] e_k q_j + \left[d_1 c_4 + d_2 c_4 + d_4 c_1 + d_4 c_3 \right] q_k e_j, \tag{31}$$

after which one can write the system of equations for unknown coefficients:

$$\begin{cases} d_1 c_1 = 1; \\ d_1 c_2 + d_2 c_1 + d_2 c_2 + d_4 c_4 q^2 = 0; \\ d_1 c_3 + d_3 c_1 + d_3 c_3 + d_4 c_4 q^2 = 0; \\ d_3 c_4 + d_4 c_2 - d_2 c_4 - d_4 c_3 = 0. \end{cases} \tag{32}$$

It is easy to solve this system:

$$d_1 = \frac{1}{c_1};$$

$$d_2 = -\frac{c_1 c_2 + c_2 c_3 - q^2 c_4^2}{c_1 \left(c_1^2 + c_1 c_2 + c_1 c_3 + c_2 c_3 - q^2 c_4^2 \right)};$$

$$d_3 = -\frac{c_1 c_3 + c_2 c_3 - q^2 c_4^2}{c_1 \left(c_1^2 + c_1 c_2 + c_1 c_3 + c_2 c_3 - q^2 c_4^2 \right)}; \tag{33}$$

$$d_4 = -\frac{c_4}{c_1^2 + c_1 c_2 + c_1 c_3 + c_2 c_3 - q^2 c_4^2}.$$

According to equations (18) and (28) the coefficients obtained determine the dielectric permittivity.

3. Discussion

The results obtained here describe the dielectric properties of semi-infinite artificial periodic structure, which consists of particles. These particles can be of different nature: atoms, molecules, nanoparticles, quantum dots, etc. equations (16) - (18) and (28), (33) describe this structure in the transparency band of the optical frequency range in the dipole approximation. The expression for the dielectric permittivity is obtained taking into account the spatial dispersion.

Acknowledgments

This work was supported by the Ministry of Education and Science of the Russian Federation, the project 3.1110.2014/K and the Competitiveness Program of National Research Nuclear University MEPhI.

References

[1]   Zhang Z and Satpathy S 1990 *Phys. Rev. Lett.* **65** 2650
[2]   Crisostomo J, Costa W A and Giarola 1993 A J *IEEE Transactions on Antennas and Propagation* **41** 1432–8
[3]   Yeh P, Yariv A. and Hong C-S 1977 *J. Opt. Soc. Am.* **67** 423-38
[4]   Koschny Th, Markoš P, Economou E N, Smith D R, Vier D C and Soukoulis C M 2015 *Phys. Rev. B* **71** 245105
[5]   Joannopoulos J D, Villeneuve P R and Fan S 1997 *Nature* **386** 143-9
[6]   Anokhin M N, Tishchenko A A and Strikhanov M N 2015 *J. Phys.: Conf. Ser.* **643** 012066
[7]   Ryazanov M I 1996 *JETP* **83** 529
[8]   Anokhin M N, Tishchenko A A, Ryazanov M I and Strikhanov M N 2014 *J. Phys.: Conf. Ser.* **541** 012023
[9]   Anokhin M N, Tishchenko A A and Strikhanov M N 2015 *PIERS Proceedings* 1354-6.
[10]   Ryazanov M I and Tishchenko A A 2006 *JETP* **103** 539

Aberration influenced generation of rotating two-lobe light fields

S P Kotova[1,2], N N Losevsky[1], D V Prokopova[1,2], S A Samagin[1], V G Volostnikov[1,2] and E N Vorontsov[1]

[1] Lebedev Physical Institute, Samara Branch, 443011, Samara, Russia
[2] Samara National Research University, 443086, Samara, Russia

E-mail: kotova@fian.smr.ru, losevsky@fian.smr.ru, prokopovadv@gmail.com, samagin@fian.smr.ru, coherent@fian.smr.ru, vorontsov2005@rambler.ru

Abstract. The influence of aberrations on light fields with a rotating intensity distribution is considered. Light fields were generated with the phase masks developed using the theory of spiral beam optics. The effects of basic aberrations, such as spherical aberration, astigmatism and coma are studied. The experimental implementation of the fields was achieved with the assistance of a liquid crystal spatial light modulator HOLOEYE HEO-1080P, operating in reflection mode. The results of mathematical modelling and experiments have been qualitatively compared.

1. Introduction

A problem of an adequate method choice for the precise determination of the location of single fluorescent molecules, quantum spots and other nanostructures arises under investigation of the substance physicochemical properties [1, 2]. And the radiating source coordinates can be changed in time. In particular these methods are of a high demand in science of materials since they permit to study the substance physicochemical properties at a level of single molecules. A precise 3D localization of nano-particles is nowadays an urgent task. Still unsolved is the problem of the precise determination of the longitudinal coordinate of the radiating nano-object in microscopy. Within the solution of this problem a special attention is paid to the use of the light fields where the structure of the intensity distribution changes with changing of, the depth of occurrence. Among these light fields, the two-lobe fields with the intensity distribution in the form of two maxima are the most promissory owing to the simplicity of the shape of their intensity distribution.

We should clarify what the main point is of the method for estimation of the longitudinal coordinate of a nanosize radiating object in a fluorescent microscope by means of the light fields with the intensity rotation. Let a sample of a certain thickness have some nanosize radiating particles (e.g. quantum spots) located at different depth of the sample. The radiation of the nanosize fluorescent objects within the microscope object domain is going through the microscope optical system into a special phase filter. The filter is a diffraction optical element that transforms the field emitted by the nanosize particle to a two-lobe light field. Provided that the radiation emitted by the particle has been transformed by the filter, then two intensity maxima appear within the microscope image plane. And their mutual orientation will determine the longitudinal coordinate of a radiation source [3-6].

Content from this work may be used under the terms of the Creative Commons Attribution 3.0 licence. Any further distribution of this work must maintain attribution to the author(s) and the title of the work, journal citation and DOI.

Published under licence by IOP Publishing Ltd

Within the present research we study the two-lobe light fields with the intensity rotation, developed using the theory of spiral beam optics. Specifically the effect exerted by basic aberrations (spherical aberration, coma, astigmatism) on the light fields under consideration is evaluated.

1.1. Two-lobe light fields description from the viewpoint of the spiral beams optics

Discovered in 90-ies of the XX century, the spiral beams have been thoroughly studied by now [7-8]. They are structurally stable under a free propagation within the scale and rotation. It was proved that the spiral beams can be formed in a way allowing generation of the field with a specified intensity distribution and desired rotation parameters of the field under its propagation. The expression that determines the rotation angle θ of the spiral beam intensity has the form:

$$\theta(z) = \theta_0 \arctan\left(\frac{2z}{k\rho^2}\right), \tag{1}$$

where k is wave number, ρ is parameter describing the beam size lateral dimension, θ_0 is parameter determining the beam rotation speed under propagation. The full rotation angle of the beam in the Fresnel zone is:

$$\theta = \theta_0 \frac{\pi}{2}. \tag{2}$$

One more property of the spiral beam is that within the initial plane it can be represented as as an expansion in Laguerre-Gaussian modes:

$$F(x,y) = \sum_{n=0}^{\infty} \sum_{m=-\infty}^{\infty} c_{nm} LG_{n,m}\left(\frac{x}{\rho}, \frac{y}{\rho}\right), \tag{3}$$

while the indices of these modes satisfy the following condition:

$$2n + |m| + \theta_0 m = const, \tag{4}$$

Thus on selecting the beam rotation parameter, θ_0 we specify the beam rotation speed and determine its functional form.

Within the present research the two-lobe light fields will be studied in the form of Laguerre-Gauss modes superposition with the rotation parameter $\theta_0 = -2$ and $\theta_0 = -4$ [9]. To calculate the phase element forming the two-lobe light field with the rotation parameter $\theta_0 = -2$, the algorithm described in [6] was modified. The Laguerre-Gauss modes superposition was used as zero-order approximation:

$$F = LG_{0,0} + LG_{1,2} + LG_{2,4} + LG_{3,6} + LG_{4,8}, \tag{5}$$

Further on the updating of the intensity distribution of the calculated field is fulfilled in order to discern two operating maxima in the intensity distribution and to reduce noises between them. The adjustment procedure consists in carrying out of two Fresnel transforms, direct and inverse, between N reference planes. One iteration of the adjustment procedure of the intensity distribution consists of N direct Fresnel transforms into each of the selected planes, and N inverse transforms into the phase element plane. We fulfilled nine such transforms, N=9. After the direct Fresnel transform, the intensity of the calculated field is changed in accordance with the following rule: if the intensity in the given point is higher than $0.5I_{max}$, where I_{max} is the maximum intensity within the selected plane (i.e. the plane where the Fresnel transform is made to), then the field amplitude remains unchanged. If the condition is not satisfied then the field amplitude in the given point becomes twice lower. The iteration procedure is stopped when the increment of the diffraction effectiveness after the i-th iteration is less than 5 %.

The phase filter for generation of the two-lobe light field with the rotation parameter $\theta_0 = -4$, was obtained when the spiral beam used had the form:

$$F = LG_{0,0} + LG_{3,2} + LG_{6,4},\tag{6}$$

The phase distributions in the shade of grey gradations and intensity distributions in the phase filter plane are shown in figure 1.

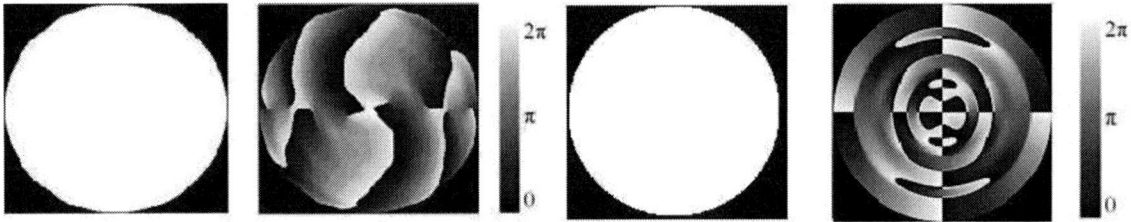

Figure 1. Intensity and phase distributions in the phase filter plane for the fields obtained by means of the development by Laguerre-Gauss modes with the rotation parameter $\theta_0 = -2$ (left) and $\theta_0 = -4$ (right).

1.2. Experimental procedure for the two-lobe light fields generation
Experiments on generating the two-lobe light fields were carried out with the aid of a phase spatial liquid-crystal modulator HOLOEYE HEO-1080. An expanded, collimated, intensity-homogeneous beam from the solid-state laser (power up to 50 mWt, λ=532 nm) was directed to the modulator HOLOEYE HEO-1080 where the calculated phase distributions of the studied light field was formed. After the light diffraction in the modulator the investigated field was formed which was then focused by the lens inscribed into the phase distribution (in the modulator). The focal length of that lens is 30 cm. The analysis of intensity distribution of the investigated fields in various cross-sections was assisted with the horizontal microscope and digital camera. In figure 2 the experimental set-up layout is shown.

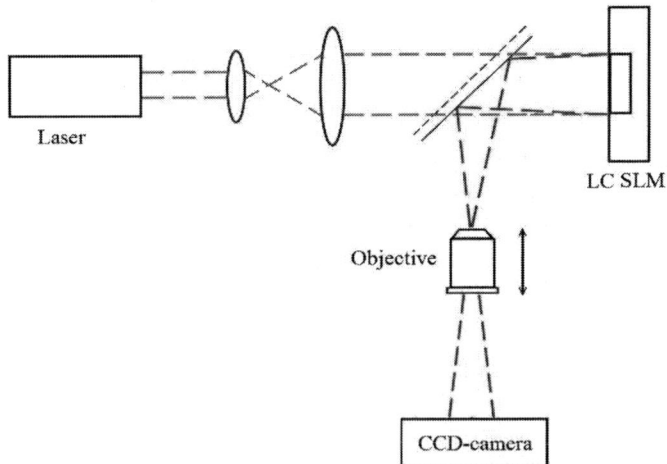

Figure 2. Layout of the experimental set-up.

The developed experimental set-up affords to study cross-sections of the light fields under investigation, perpendicular to the axis of their propagation. The studied field's structure can be judged by the intensity distributions obtained in different sections. Typical patterns of the intensity distribution for the two-lobe light fields are illustrated in figure 3.

CORSCS2015   IOP Publishing
Journal of Physics: Conference Series **740** (2016) 012013   doi:10.1088/1742-6596/740/1/012013

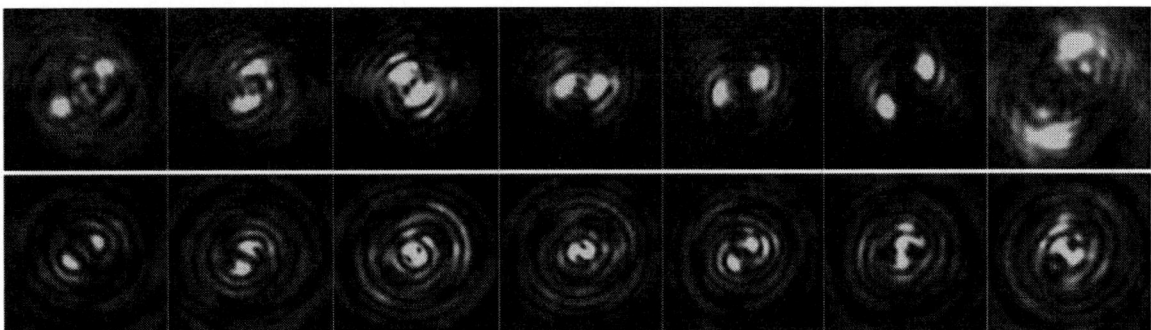

Figure 3. Intensity distribution in sections perpendicular to the axis of propagation of the two-lobe light fields with the rotation parameter $\theta_0 = -2$ (upper row) and $\theta_0 = -4$ (lower row).

The information on the spatial location of the maxima of intensity can be extracted from the experimentally obtained data. First the angle of rotation of the two maxima from the initial state (horizontal line) is measured. Each value obtained for the rotation angle of the intensity distribution has its corresponding distance from the focusing plane to the beam section where this intensity distribution is registered. With the data obtained it is possible to form the relationship between the angle of rotation and the distance to the beam waist. The rotation of the intensity distribution under free spatial propagation is fulfilled counterclockwise (figure 3). Figure 4 shows the experimental points and diagrams (1) for the rotation parameters $\theta_0 = -2$ and $\theta_0 = -4$. It is seen that the experimentally obtained dependencies are well matching the analytical ones.

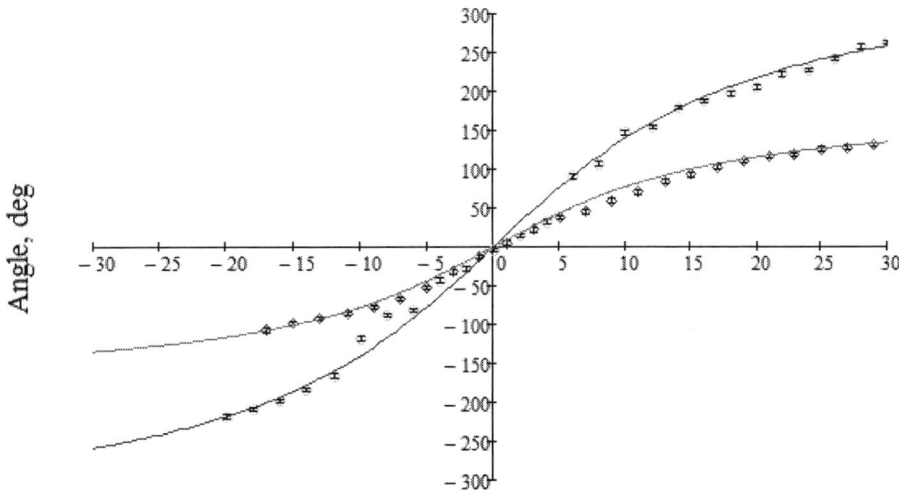

Figure 4. Rotation angle θ of the intensity distribution versus distance from the focusing plane diagram. Red line is the analytical dependency for $\theta_0 = -2$; blue line – for $\theta_0 = -4$. Error of measurement is $5°$.

87

2. Aberrations impact on the two-lobe light fields. Mathematical modeling and experimental data.

While dealing with real optical systems one is inevitably facing the aberrations existing in them. Besides, the radiating nano-particles are located inside a studied sample where some heterogeneities causing the phase distortions can be found. These facts testify that then investigation of the aberrations influence on two-lobe light fields is an important problem. This section of the present work is dedicated to the study on the aberrations influence on a two-lobe light field.

2.1. Simulation of the aberrations influence on two-lobe light fields generation

The aberrations were specified in the representation of the Zernike polynomials [10], in accordance with the OSA standard normalization [11].

$$Z_n^m(r,\theta) = \begin{cases} N_n^m R_n^m(r)\cos(m\theta), & for\ m \geq 0; \\ -N_n^m R_n^m(r)\sin(m\theta), & for\ m < 0. \end{cases} \tag{7}$$

$$R_n^m(r) = \sum_{s=0}^{(n-|m|)/2} \frac{(-1)^s (n-s)!}{s!\left(\frac{n+|m|}{2}-s\right)!\left(\frac{n-|m|}{2}-s\right)!} r^{n-2s} ; \tag{8}$$

$$N_n^m = \sqrt{\frac{2(n+1)}{1+\delta_{m0}}} ; \tag{9}$$

where r and θ are polar coordinates.

The computational modeling was carried out to simulate the influence of astigmatism (Z_2^2), coma (Z_3^1) and spherical aberrations (Z_4^0) on the studied light fields with the rotation of the intensity distribution. The phase transmission of the calculated mask was multiplied by the phase transmission adequate to the specified aberration type. The aberration value ranged from $\lambda/32$ to λ. Then the intensity distributions were calculated for the sections perpendicular to the axis of the light field propagation within the area of focusing by a lens with the focal distance of 300 mm, in cross sections spaced one from another by 1mm. It was revealed that when the value of the spherical aberration of $\lambda/16$, coma $\lambda/16$ and astigmatism $\lambda/8$ in the intensity distribution of the light field with the rotation parameter $\theta_0 = -2$ it is still possible to discern two spots with the maximum illumination. Though for the astigmatism of $\lambda/8$ there is an area inside the range where the two spots are not clearly cut. The simulation results for the mentioned cases are given in figure 5. For the light field with the rotation parameter $\theta_0 = -4$ the aberration value allowing to discern two maxima of the intensity distribution is $\lambda/16$. This result is illustrated in figure 6. It should also be mentioned that with the rise of the aberration value, the area where the two maxima of the intensity distribution can be discerned, is reducing.

Table 1 contains the data obtained by the computational modeling results processing. This data confirm that the rotation of two well-separated intensity maxima can be observed under low aberration values. The problem of evaluating the depth of occurrence of the radiating nano-object requires that two-lobe intensity distribution rotated by 180°. On assuming this condition we'll consider the aberration influence insignificant provided that the angle of intensity distribution rotation is 180°. By using table 1 and on assuming this condition we can conclude that the aberrations influence is nonessential for the value of $\lambda/16$, both for the field with the rotation parameter $\theta_0 = -2$ and for field with the rotation parameter $\theta_0 = -4$.

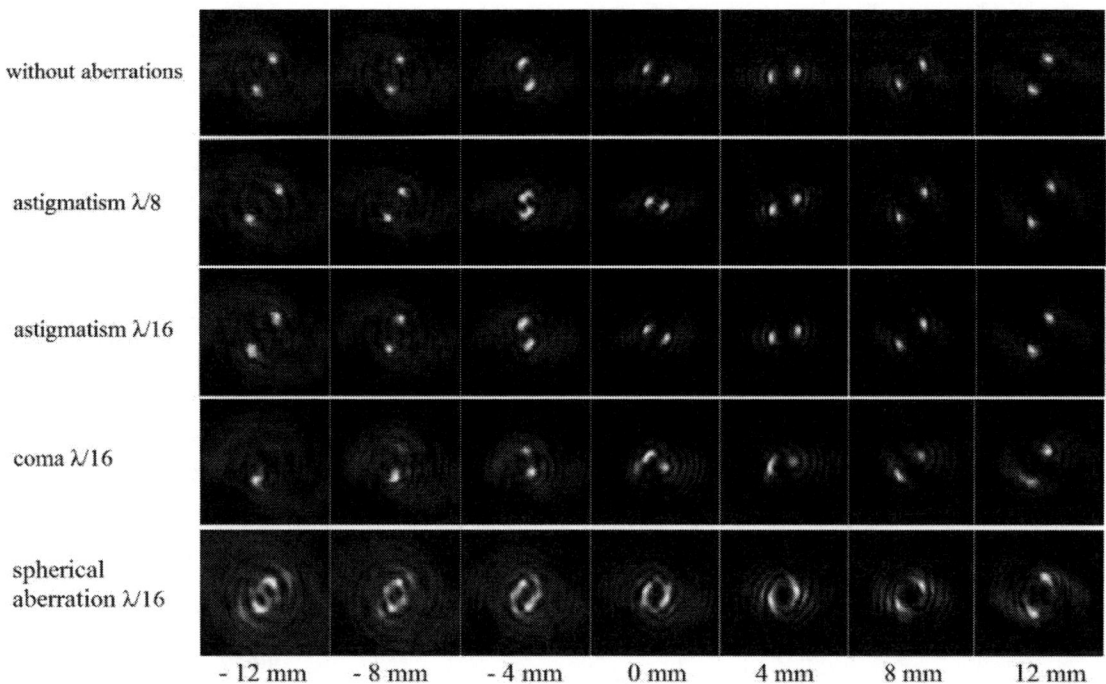

Figure 5. Results of aberrations influence modeling for the light field with the rotation parameter $\theta_0 = -2$.

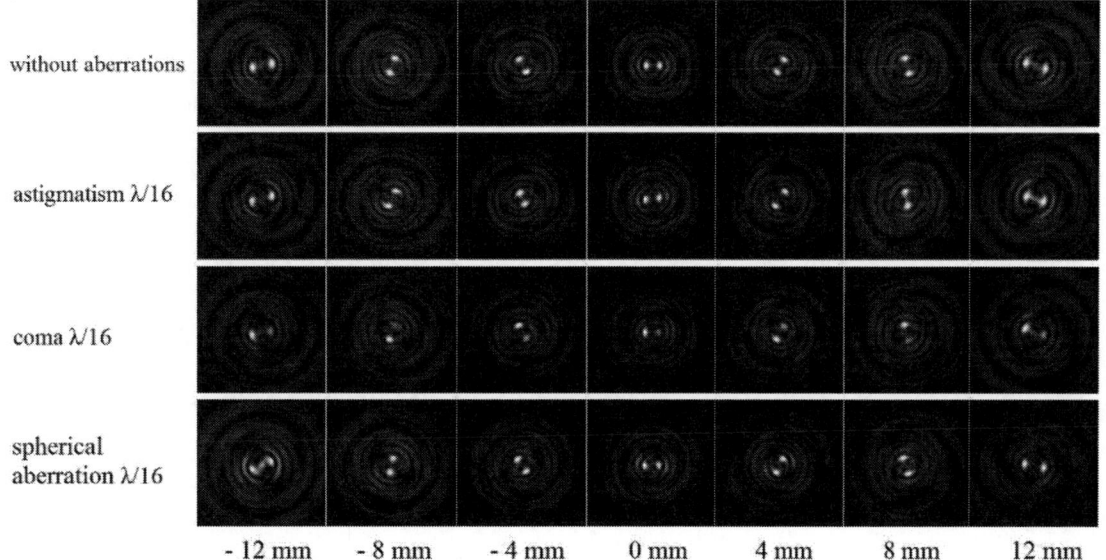

Figure 6. Results of aberrations influence modeling for the light field with the rotation parameter $\theta_0 = -4$.

Table 1. Full rotation angle of two maxima in the intensity distribution of the light fields being researched under aberrations.

The light field with the rotation parameter $\theta_0=-2$			The light field with the rotation parameter $\theta_0=-4$		
Astigmatism	$\lambda/32$	214°	Astigmatism	$\lambda/32$	370°
	$\lambda/16$	212°		$\lambda/16$	285°
	$\lambda/8$	30°, 98°		$\lambda/8$	250°
	$\lambda/4$	-		$\lambda/4$	-
	$\lambda/2$	-		$\lambda/2$	-
	λ	-		λ	-
Coma	$\lambda/32$	208°	Coma	$\lambda/32$	370°
	$\lambda/16$	185°		$\lambda/16$	372°
	$\lambda/8$	143°		$\lambda/8$	15°
	$\lambda/4$	-		$\lambda/4$	-
	$\lambda/2$	-		$\lambda/2$	-
	λ	-		λ	-
Spherical aberration	$\lambda/32$	223°	Spherical aberration	$\lambda/32$	365°
	$\lambda/16$	216°		$\lambda/16$	291°
	$\lambda/8$	20°		$\lambda/8$	-
	$\lambda/4$	-		$\lambda/4$	-
	$\lambda/2$	-		$\lambda/2$	-
	λ	-		λ	-

2.2. Experimental study on the aberrations influence on two-lobe light fields

A number of experiments on the two-lobe light fields generation under aberrations (spherical aberration, coma, astigmatism) of various values ($\lambda/32$, $\lambda/16$, $\lambda/8$, $\lambda/4$, $\lambda/2$, λ) in accordance with the OSA standard normalization. Under the experiments the phase distributions formed by the spatial light modulator represents a superposition of the phase distribution of the studied field and the initial aberration of a certain value. The aberrations were specified in the representation of the Zernike polynomials in the form: Z_2^2 – astigmatism, Z_3^1 – coma, Z_4^0 – spherical aberration. The variable r was ranging within [0, 1]. During the phase mask calculation the scaling of radial coordinate was carried out, so that the aperture radius was equal to 1.

The registration of the obtained field intensity distribution was fulfilled within the planes perpendicular to the axis of the light field propagation. In the course of a visual analysis of the experimentally obtained intensity distributions of the researched light fields under aberrations it was revealed that at the value $\lambda/16$ of spherical aberration, coma and astigmatism it is feasible to extract, basing on the intensity distribution, the information on the maxima location for the considered fields. This can be clearly seen in figures 7 and 8. The data obtained are coinciding with the mathematical modeling results for the initial aberrations influence on the light fields under investigation.

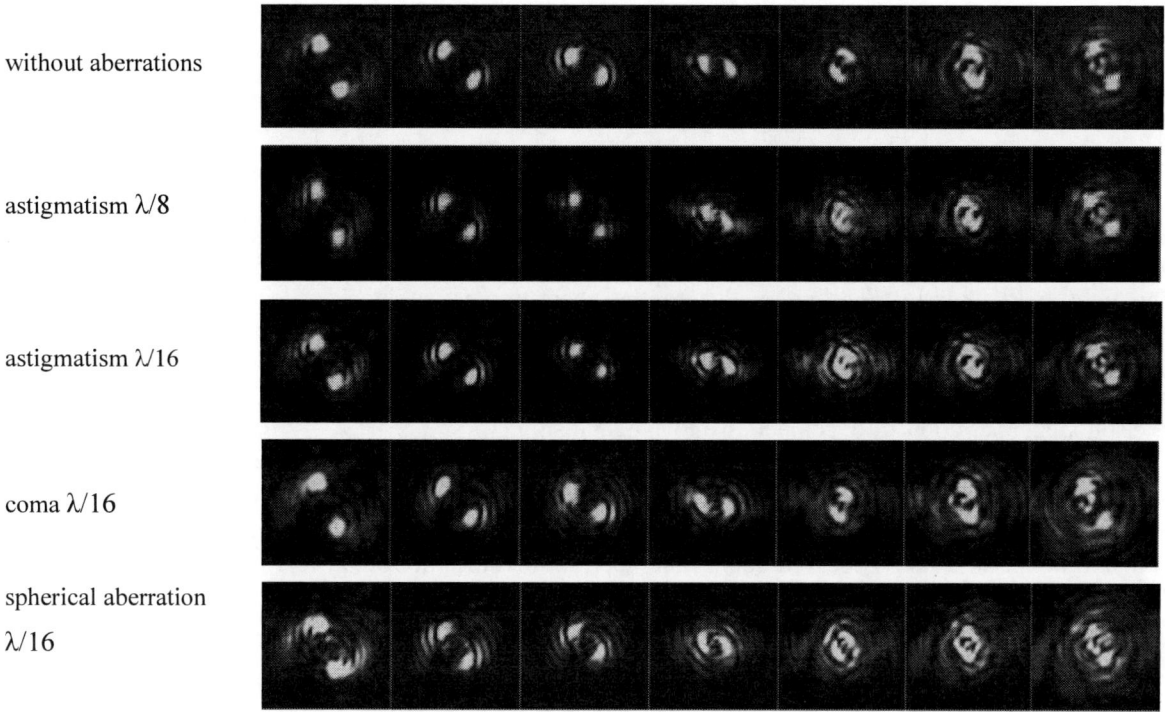

Figure 7. Results of experiments on the aberrations influence on the light field with the rotation parameter $\theta_0 = -2$

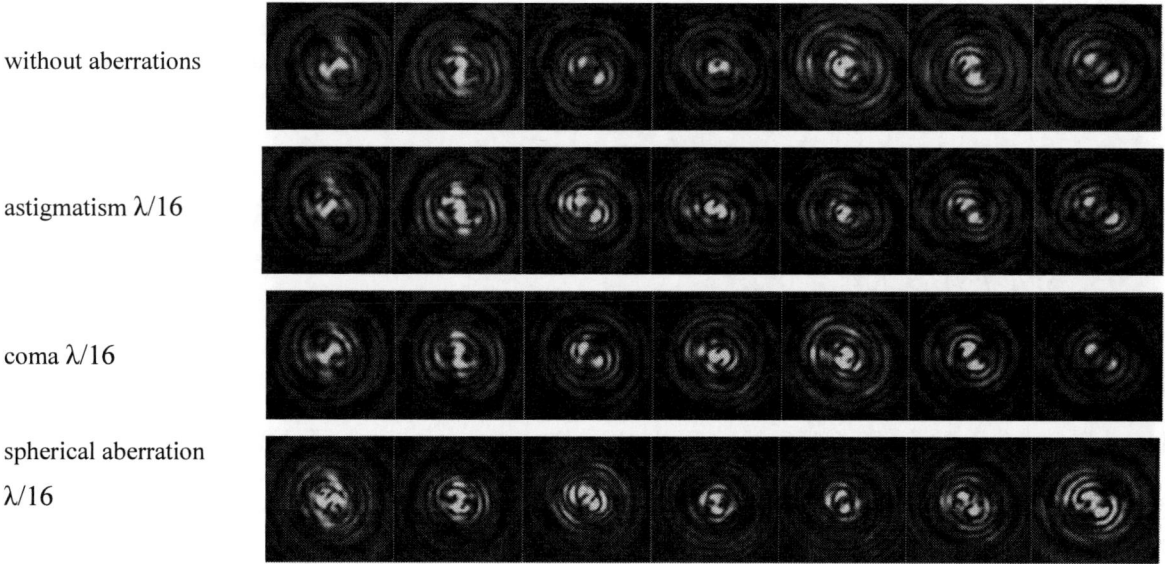

Figure 8. Results of experiments on the aberrations influence on the light field with the rotation parameter $\theta_0 = -4$

3. Results

The paper presents the results obtained for the generation of two-lobe light fields with the rotation of intensity distributions. The light fields have been obtained on the basis of the superposition of Laguerre-Gaussian modes, with the rotation parameters $\theta_0 = -2$ и $\theta_0 = -4$. The experimentally obtained relationships of the rotation angle and the distance agree well with the theoretic ones (figure 4).

First studies of the aberrations (astigmatism, coma, spherical aberration) influence on the generation of two-lobe light fields have been fulfilled. The computational modeling of the aberrations influence showed that with the rise of the aberration value the range of rotation angles of the two discernible maxima in the intensity distribution is decreased (table 1). It is caused by the distortion of the light field intensity distribution due to the aberration value increase, as a result it becomes impossible to discern the two maxima. A quantitative analysis of the numerical and life-size modeling permits to conclude that from the intensity distribution pattern, the information on the angular location of the maxima can be obtained for the studied fields with the rotation parameter $\theta_0 = -2$ and $\theta_0 = -4$ at the spherical aberration $\lambda/16$, coma $\lambda/16$ and astigmatism $\lambda/16$.

Note that normalization of the Zernike polynomials by the OSA standards and that used in [9] are different. In order to transfer to the aberrations described in [9] one should multiply the astigmatism value under the OSA standard by $\sqrt{6}$, coma – by $2\sqrt{2}$, astigmatism – by $\sqrt{10}$. On making this transition experimentally observed admissible aberration values are obtained for astigmatism $0,153\lambda$, for coma $0,177\lambda$ and for spherical aberration – $0,198\lambda$.

Acknowledgements

The research project was supported by the Program for Basic Researchers of the Physical Sciences Department of RAS "Fundamentals and experimental realization of promising semiconductor lasers in the interests of industry and technology", by the RFBR Project No. 16-29-11809, by the Educational and Scientific Complex of the Lebedev Physical Institute and by Competitiveness Program of National Research Nuclear University MEPhI.

References

[1] Naumov A V 2013 Low temperature spectroscopy of organic molecules in solid matrices: from the Shpolsky effect to the laser luminescent spectromicroscopy for all effectively emitting single molecules *Phys. Usp.* **56** 605–22

[2] Naumov A V, Eremchev I Y and Gorshelev A A 2014 Laser selective spectromicroscopy of myriad single molecules: tool for far-field multicolour materials nanodiagnostics *Eur. Phys. J. D* Colloquium paper

[3] Lew M, Thompson M, Badieirostami M and Moerner W 2010 In vivo three-dimensional superresolution fluorescence tracking using a double-helix point spread function *Proc. of SPIE* **7571** 75710Z

[4] Backlund M at al 2013 The double-helix point spread function enables precise and accurate measurement of 3D single-molecule localization and orientation *Proc. of SPIE* **8590** 85900L.

[5] Grover G and Piestun R 2013 New approach to double-helix point spread function design for 3D super-resolution microscopy *Proc. of SPIE* **8590** 85900M

[6] Pavani S R P, and Piestun R 2008 High-efficiency rotating point spread functions *Opt. Express* **16**(5) 3484–9

[7] Abramochkin E and Volostnikov V 1993 Spiral-type beams *Optics Communications* **102**(3) 336

[8] Abramochkin E G and Volostnikov V G 2010 *Modern Optics of Gaussian beams* (FIZMATLIT, Moscow)

[9] Volostnikov V G, Vorontsov E N, Kotova S P, Losevskiy N N and Prokopova D V 2015 A Diffraction Element Used to Evaluate the Depth of Bedding of Nano-Sized Radiating Objects *EPJ Web of Conferences* **103** 10007

[10] Born M and Wolf E 1980 *Principles of Optics* (Pergamon Press, Oxford)

[11] Zernike Polinomials (OSA standard). Source: ⟨https:// www.optics.ru/info/tops4.pdf⟩

CORSCS2015   IOP Publishing
Journal of Physics: Conference Series **740** (2016) 012014   doi:10.1088/1742-6596/740/1/012014

Radioprotective Action of Low-Intensity Light into the Red Absorption Band of Endogenous Molecular Oxygen

A V Ivanov[1], A A Mashalov[1] and S D Zakharov[2,3]

[1] Blokhin Russian Cancer Research Center, 115478, Moscow, Russia
[2] Division of Quantum Radiophysics, Lebedev Physical Institute, Russian Academy of Sciences, 119991, Moscow, Russia
[3] National Research Nuclear University MEPhI, 115409, Moscow, Russia

E-mail: ivavi@yandex.ru, imhaemo_6june@ronc.ru, stzakhar@sci.lebedev.ru

Abstract. Application of ionizing radiation in oncology (radiation therapy) is a widespread way to eliminate malignant tumors. Normal tissues are inevitable included in any radiation field, and their reliable protection is actual till now. All attempts to solve the problem are based on search of effective radioprotectors, i.e. chemical compounds of various classes, which should be entered into the patient. To date about 50,000 compounds with some radioprotection properties had been tested, but the most effective of them have been simultaneously the most toxic. Here the preliminary results of researches devoted to development of an optical technique on basis of the light-oxygen effect for the protection of women with breast cancer from side effects of the radiation therapy are presented. A low intensity emission of the semiconductor laser in a red spectral interval was used to excite a very small quantity of endogenous molecular oxygen in $O_2(^1\Delta g)$ state. It is shown, that application of the method at occurrence of earliest signs of radiation injury allows notably reducing dangerous breaks in radiation therapy course.

1. Introduction

Basic tendency in oncology is reduction of radiotherapy course duration by means of increasing the ionizing radiation intensity. However side effects, i.e. radiation induced injuries of normal tissues, forces sometimes to interrupt the standard course to give to patient a time for their restoration. Such interrupts are extremely undesirable as they allow to some malignant cells is adapted and increase their radioresistance [1].

This global problem is not completely solved till now. A well-known strategy consists in search of antioxidants with low toxicity which would be capable to neutralize effectively radicals induced by the ionizing radiation [2]. Other ways, such as hyperbaric oxygenation, hyperthermia, hyperglycemia, have been offered also, however those have not found wide clinical application.

Here we report on possibility to increase resistance of women with breast cancer to post-radiation damages by help of a spectral-selective light. A fundamental basis of the technique is the light-oxygen effect (LOE), universal phenomenon found out at all levels of the biological organization - from pure water and protein solution up to human [3-5].

The primary photophysical act of the light-oxygen activation (LOT) is direct, i.e. without any photosensitizers, excitation of blood-dissolved oxygen molecules by photons into a metastabil singlet energetic state [6]. At small doses of the singlet oxygen an activation of patient's aerobic metabolism

Content from this work may be used under the terms of the Creative Commons Attribution 3.0 licence. Any further distribution of this work must maintain attribution to the author(s) and the title of the work, journal citation and DOI.
Published under licence by IOP Publishing Ltd

occurs, whereas large doses of this agent, on the contrary, have caused cell damages. Wavelength of light source, its power and exposition dose are interdependent and must be selected carefully as molecular oxygen absorbs light only in narrow spectral intervals including a "red" band. Semiconductors lasers are convenient instruments for such purposes [7].

2. Materials and Methods

Light induced radioprotection effect was studied for women with breast cancer. Two group of 26 women (non-randomized research) and 37 women (randomized research) with a control (without light activation) of 30 women only for second group were formed. All patients have been subjected breast therapy with total dose up to 62 Gy after surgical intervention and/or 2-4 courses of polychemotherapy. Patients received also the standard pharmacological preparations warning skin's radiation injuries.

The 635 nm wavelength semiconductor laser was used for the LOT. The mammary gland and axillary's area were subjected light irradiation. A LOT course has consisted of 12-15 daily procedures 1 hour before of each radiation impact. The light power density on patient's skin and total exposition dose were 12-15 mW/cm^2 and about 3 J/cm^2 for one procedure correspondently.

3. Results

In first group the LOT procedures have been begun with delay when total dose of 28 - 44 Gy has been supplied and most of patients had a skin's radiation pathology of II - III degrees. Nevertheless, at 21 from 26 women a medical effect has been received. The effect was expressed in the form of reduction or disappearance of pain, itch, a hypostasis and hyperemia in an irradiated zone of skin. It has allowed avoiding undesirable breaks in treatment. The others five patients have been forced to interrupt the course on 8 - 10 days for treatment of excessive skin injuries; note that the LOT did not stop during those breaks.

In other group of 37 patients the LOT has been applied at once after occurrence of first signs of radiation injury (erythemas, radioepithelites). The LOT has proceeded up to the termination of the radiotheraphy course. Results are summarized in the Table.

Table. Light induced reduction of skin's radiation injuries at radiotherapy of breast cancer

Operation	Total number of patients	Number of patients with different grade of radiotoxicity				Breaks in treatment (from 8 to 12 days)
		O	I	II	III	
Belated light therapy	26	-	4	16	6	5 (19%)
Timely light therapy	37	2	25	8	2	0
Control (without light)	30	-	12	14	4	8 (27%)

4. Discussion

It has long been known that the oxygen molecules in electron excited state, named the singlet oxygen, can damage cells. In oncology, singlet oxygen is produced by means of dyes (photosensitizers) and is used in the photodynamic therapy of cancer. However, we have showed earlier that the cell response to photodynamic action depends on the singlet oxygen dose and formation rate and varies from cell stimulation up to cell destruction. Moreover, both cellular activation and cellular destruction can be realized without photosensitizers using direct photoexcitation of endogenous oxygen. Surprisingly, but from the lasers used for this purpose, large light fluxes are not required. This phenomenon is the light oxygen effect.

Here we show other surprising property of the LOE. Using it, it is possible to increase a protection of healthy cells against damaging action of an ionizing radiation, not reducing its injuring potential in relation to malignant cells. This way of increase of efficiency of radiotheraphy seems to us promising. Its advantage over other methods is that it does not require entry into a patient's blood of any exogenous drugs: oxygen is abundant in tissues. Also, the technique is non-toxic.

Feature of singlet oxygen in small doses have activating effect on organism and damaging effects in large doses is apparently the manifestation of a general phenomenon called hormesis. Many medicals exhibit the hormesis effect, e.g. curare, poison in large doses and drug in low ones. The difficulty is to find a dose boundary separating opposing responses. In our context, the optimal photon dose is critically dependent on the wavelength. More details on this issue will be presented elsewhere.

5. Conclusion
Protection of normal tissues from ionizing radiation is the most serious problem in radiation oncology. The light-oxygen effect confirmed earlier at all levels of the biological organization - from pure water and protein solution up to human can be used for such purposes. The molecular oxygen dissolved in blood is a primary target for light photons whereas singlet oxygen, the product of this interaction, is an activating agent. Optimum results are reached when the light protection procedures is followed earliest signs of tissue damage induced by ionizing radiation.

Acknowledgements
This work was supported by Competitiveness Program of National Research Nuclear University MEPhI.

References
[1]   Miller D L, Balter S, Noonan P T and Georgia J D 2002 *Radiology* **225** 329
[2]   Shindo Y, Witt E, Han D, Epstain W and Packer L 1994 *J. Investig. Dermatol.* **102** 122
[3]   Zakharov S D, Eremeev B V and Perov S N 1989 *Soviet physics – Lebedev Institute Reports* **1** 19
[4]   Zakharov S D and Ivanov A V 1999 *Quant. Electr.* **29** 1031
[5]   Zakharov S D and Ivanov A V 2005 *Biophysics* **50**(Suppl. 1) S64
[6]   Minaev B F, Murugan N A and Agren H 2013 *Int. J. Quant. Chem.* **113** 1847
[7]   Zakharov S D, Korochkin I M, Yusupov A S, Bezotosnyi V V, Cheshev E A and Frantzen F 2014 *Semiconductors* **48** 129

Transmission of large amounts of scientific data using laser technology

E A Isaev[1,2] and P A Tarasov[3]

[1] National Research University «Higher School of Economics», 101000, Moscow, Russia
[2] Pushchino Radio Astronomy Observatory of Astro Space Center of P.N. Lebedev Physics Institute, RAS, 142290, Pushchino, Russia
[3] Institute of Mathematical Problems of Biology, RAS, 142290, Pushchino, Russia

E-mail: is@itaec.ru, tpahan@yandex.ru

Abstract. Currently, the volume of figures generated by different research scientific projects (the Large Hadron Collider (Large Hadron Collider, LHC), The Square Kilometre Array (SKA)), can reach tens of petabytes per day. The only technical solution that allows you to transfer such large amounts of scientific data to the places of their processing is the transfer of information by means of laser technology, using different propagation environment. This article discusses the possibility of data transmission via fiber-optic networks, data transmission using the modulation binary stream of light source by a special LED light source, the neccessity to apply laser technologies for deep space communications, the principle for an unlimited expansion of the capacity of laser data link. Also in this study is shown the need for a substantial increase in data transfer speed via a pre-existing communication networks and via the construction of new channels of communication that will cope with the transfer of very large scale data volumes, taking into account the projected rate of growth.

1. Introduction

Recently, the rapid development of various scientific fields (such as radio astronomy, bioinformatics, climatology) has lead to appearance of scientific projects, which form massive volumes of the order of tens of terabytes of figures in a short time period. This leads to an extreme increase of the volume of transmitted information in data networks. Therefore, we are faced with the necessity of the solution of the problem of a strong increase of the volume of figures transmitted in the local and regional data networks in contemporary world. It already causes the exhaustion of available resources, and the real needs of requirements indicate the continuation of the growth of information flows tenfold and hundredfold in some cases [1].

2. Current situation

Let's consider some of the largest Russian and foreign research projects that operate with massive data arrays.

Nowadays the radio telescope RT-22 (figure 1) is used as a tracking station for the project "Radiostron" of Physical Institute Astro Space Center. PN Lebedev (ASC FIAN). The goal of this project is to conduct scientific observations using radio astronomy telescope, which was established in the scientific-production association Lavochkin [2] and is mounted on the spacecraft "Spektr-R". The daily stream of figures from this radio astronomy center is just over 1.28 terrabytes. This project uses communication channel of 1Gbps bandwidth for its operations, which connects the tracking station and a buffer data center, and located on the territory of the Pushchino Radio Astronomy Observatory, and the center of the processing of scientific information Astro Space Center in Moscow [3].

Figure 1. Radiotelescope RT-22 [4].

Research related to the human genome, provide a wealth of information also. For example, the data center of the Beijing Genomic Institute, consisting of more than 500 nodes, each day produces about 10 terabytes of "raw" data [5].

Fluorescence microscope IsoView, created in the Howard Hughes Medical Institute, is able to visualize the cellular dynamics in all three dimensions in large organisms with a resolution of up to the definition of individual cells structure. A typical experiment fluorescence microscope lasts more than one hour. Four cameras transmit approximately 3.2 gigabytes of data per second, resulting in about 10 terabytes of raw data [6].

Another project that generating huge amounts of data, is an array of telescopes, LOFAR (LOw Frequency ARray - «low-frequency antenna array"). Radio telescope explores the low-frequency radio waves to search for the first stars and galaxies in our Universe, the potential signals of extraterrestrial intelligence, as well as studying black holes and pulsars. The project connects about 20,000 radio antennas located throughout Europe. LOFAR generates about 138 terabytes of data per day [7].

Nowadays for the high-speed transmission of this vast amounts of information are widely used laser technology, where as a carrier information signal, the electromagnetic radiation in the optical range. The ubiquitous distribution in the commercial and scientific exploitation has optical fiber communication. It uses special optical cables as the transmission medium, comprising filaments of optically transparent material (glass, plastic) that serve to transfer light internally by total internal reflection. Widespread this systems received due to a number of advantages, such as high bandwidth and low attenuation of the signal, as compared to transmission systems using copper cables or radio broadcast as the transmission medium, and in days.

Existing commercial communication line speed at this time is usually no more than 10 Gbps. However, the possibilities for transmitting in scientific data networks are higher. For example, the internal rate of scientific network of the US Department of Energy's ESnet (Energy Sciences Network) is 100 Gbps [8], it will be coupled by the channels with the European scientific and educational GEANT network. Transatlantic expansion will provide users the capacity of 340 Gbps and it can serve dozens of scientific collaborations [9].

3. Investigations and possible solutions
It should be noted that the existing data transmission speeds insufficient for the growing needs of science. It is necessary to constantly upgrade the existing communication channels, communication equipment, data transmission protocols, as well as the running of new communication channels and the launch of new data transmission technologies.

The research on this issue is constantly carried out and progress has been made in this field, allowing to look optimistically into the future. Based on these results it can be concluded that a fiber-optic connection can cope with the transfer of a large amount of scientific data. Further, the article describes some of the research groups of scientific capabilities of fiber-optic communication.

In 2014, a joint team of researchers from the University of Technology Eindhoven (Eindhoven University of Technology) and the University of Central Florida (CREOL) has created a new type of fiber optic cable capable of transmitting on one strand fiberglass data at a rate of 255 Tbps [10].

In early 2014, the British company Alcatel-Lucent and British Telecom were able to achieve a data rate of 1.4 Tbps on a standard fiber optic cable that was laid quite a long time, using a set of standard communication equipment. This speed was achieved by means of a unique hardware solution - new protocol called "Flexigrid", which allows you to overlay multiple independent signals to each other and transmit it over a single optical cable [11].

And the absolute record of fiber optic data transmission speeds currently achieved by Nippon Telegraph and Telephone Corporation (NTT) Japanese companies. They worked together with three partner organizations, company Fujikura Ltd., Hokkaido University and the Danish University of Technology (Technical University of Denmark, DTU). Their experiment in September 2012 showed a record speed of information transmission over a single fiber. During the test of a new line of communication, specialists recorded data rate of 1 Ptbs over fiber optic cable with 12 channels, and a light guide length of 52.4 kilometers. This is orders of magnitude greater than the rate cables are now in commercial operation [12].

Russian company "T8" has established several world records on the fields of data rate transfer. For example, in 2014 on a sector of 500.4 km length with a channel speed of 100 Gbps was achieved speed 1Tbps due to amplifiers with remote pumped ROPA and simultaneous transmission of 10 channels of information. In 2013, the successful transmission of signals at 100 Gbps was achieved over a distance of 4,000 kilometers in the current link in the "Volga" equipment. And now T8 is developing a high-speed WDM platform (DWDM) with total capacity of 25 Tbps [13].

It should be noted that currently there exists a development of high-speed data transmission by means of laser technology in the air. For example, in 2014 a group of Russian companies "Stins Coman" introduced a wireless local area network "Beamcaster", which uses data transmission by laser beam in the air, and the data rate which can be up to 40 Gbps by virtue of this technology [14].

National Aeronautics and Space Administration (NASA) proposes to use laser beams to transmit information in a vacuum, because the wavelength of visible light is much shorter than that of radio waves, therefore so the use of the optical transmitter will transmit much more of figures. In addition, the laser is much better "hover" at the target, therefore transfer requires less energy. In the current experiment in 2013 LADEE probe (figure 2), came to a circular orbit at an altitude of 235 kilometers above the lunar surface and spent laser communication session between the Moon and the Earth by an on-board device Lunar Laser Communication Demonstration (LLCD). The authors of experiment said that the data rate at a distance of 385,000 kilometers between the Moon and the Earth was a record 20 Mbps, and between the Moon and the Earth - 622 Mbps.

Figure 2. Spaceprobe LADEE [15].

Featured data rate appeared to be several times higher than the transmission speed of the radio waves by the spaceships reached the surface of the moon ever before. Furthermore, the signal receiver in Earth have four antennas 30 cm in diameter. This size is orders less than that of the antennas receiving data over radio waves from space. In the future, NASA is planning to develop this information transmission system, and it allows you to send and receive high-definition images of probes and 3D video from deep space. Also, NASA is planning an another long-term experiment on the LCD (Laser Communications Relay Demonstration) laser communications, which will be launched in 2017 [16].

The development teams for several years experimenting with the new data transfer technology, the LEDs in the fluorescent. The project is called "light fidelity", or "li-fi". With this technology, the Mexican company Sisoft together with scientists from the Autonomous Technological Institute of Mexico in 2014 reached a data rate of 10 Gbps [17]. And at Oxford University are developing a transceiver, which receives the signal from the optical fiber, amplifies it and transmits the room at a speed of 224 Gbps by means of this technology [18]. The data rate is much higher than the maximum data transfer speeds via Wi-Fi technology.

Recently the newest technology have appeared, enabling virtually unlimited bandwidth to expand the data. We are talking about the use of one of the quantum characteristics of photons, namely their "twist" - the orbital angular momentum of photons relative to their direction of propagation. This very property helps photons to implement the transfer of information not by classical "zero" and "one", but by the qubits in a quantum superposition. This makes it possible for us to realize for the unit the transmission of information an infinite number of states, describing arbitrary points of the multidimensional space. Despite the apparent high noise immunity of this communication channel, recent experiment was carried out by a group of scientists under the guidance of professor Zeilinger [19]. This experiment has proved its high reliability even when the transfer of information id effected through the atmosphere. The experiment revealed that, despite the high number of atmospheric turbulence the quantity of incorrectly transmitted information does not exceed one percent. This allows to conclude about the high reliability of this transmission channel. Transfer rate was 4 pixels per second, but no optimization was carried out. This experiment showed that at large distances the turbulence of the atmosphere in the process of transfer has no significant impact on the reliability of data transmission. It also revealed the prospect of this method when sending large amounts of information, including for communicating with interplanetary spacecraft. In the future, announced the establishment of a channel with 11 states twist [20].

4. Conclusion

To date, the available communication links cope with the flow of large volumes of scientific data, and providing remote access of researchers to the complex scientific equipment and computing resources in real-time. However, due to the ever increasing volume of scientific information generated by the scientific projects there is the need for a fundamentally new technological solutions, using laser technology to increase the scientific data rates.

Acknowledgements

This work was supported by Competitiveness Program of National Research Nuclear University MEPhI.

References

[1]   Isaev E A, Kornilov V V and Tarasov P A 2013 Scientific computer networks - challenges and successes in organizing the exchange of large amounts of scientific data *Mathematical Biology and Bioinformatics* **8**(1) 161–81

[2]   The Radioastron project http://www.asc.rssi.ru/radioastron/index.html [Accessed 24 03 2016]

[3]   Shackaya M V, Girin I A, Isaev E A, Lihachev S F, Pimakov A S, Seliverstov S I and Fedorov N A 2012 The organization of the processing center of scientific information for radio interferometric projects *Space investigations* **50**(4) 346–50

[4]   The PRAO project http://www.prao.ru/radiotelescopes/images_new/rt22_big.jpg [Accessed 24 03 2016]

[5]   Petsko G A 2010 Rising in the East *Genome Biology* **11**

[6]   Chhetri R K, Amat F, Wan Y, Höckendorf B, Lemon W C and Keller P J 2015 Whole-animal functional and developmental imaging with isotropic spatial resolution *Nature Methods* **12** 1171–8

[7]   Mattmann C A et al 2013 Scalable Data Mining, Archiving, and Big Data Management for the Next Generation Astronomical Telescopes *Big Data Management, Technologies, and Applications* (IGI Global) chapter 9

[8]   The ESnet project https://www.es.net/engineering-services/the-network/network-maps/ [Accessed 24 03 2016]

[9]   The ESnet project https://www.es.net/news-and-publications/esnet-news/2014/esnet-extends-100g-connectivity-across-atlantic/ [Accessed 24 03 2016]

[10]   Van Uden R G H, Correa A R, Lopez E A, Huijskens F M, Xia C, Li G, Schülzgen A, de Waardt H, Koonen A M J and Okonkwo C M 2014 Ultra-high-density spatial division multiplexing with a few-mode multicore fibre *Nature Photonics* **8** 865–70

[11]   *Computerweekly* magazine http://www.computerweekly.com/news/2240212985/BT-and-Alcatel-Lucent-test-14Tbps-broadband [Accessed 24 03 2016]

[12]   The NTT Group *World Record One Petabit per Second Fiber Transmission over 50-km* http://www.ntt.co.jp/news2012/1209e/120920a.html [Accessed 24 03 2016]

[13]   The «T8» company http://t8.ru [Accessed 24 03 2016]

[14]   The group of companies «Stins coman» http://www.stinscoman.com/674.html [Accessed 24 03 2016]

[15]   NASA homepage

https://www.nasa.gov/sites/default/files/styles/ubernode_alt_horiz/public/thumbnails/image/l adee.jpg?itok=u2kBwYuP [Accessed 24 03 2016]

[16] NASA homepage Keesey L *NASA to Demonstrate Communications Via Laser Beam* http://www.nasa.gov/topics/technology/features/laser-comm.html [Accessed 24 03 2016]

[17] *Engineering and Technology* Magazine http://eandt.theiet.org/news/2014/jul/li-fi-sisoft.cfm

[18] Gomez A, Shi K, Quintana C and Sato M Beyond 100-Gb/s Indoor Wide Field-of-View Optical Wireless Communications *Photonics Technology Letters* **27**

[19] Krenn M *et al* 2014 Twisted light communication through turbulent air across Vienna *New J. Phys.* **16** 113028

[20] Krenn M *et al* 2015 Twisted photon entanglement through turbulent air across Vienna *PNAS* **112**(46)

Funds support the construction of control systems design

V Vlasov and A Tolokonsky

National Research Nuclear University MEPhI, Kashirskoe shosse 31, 115409, Moscow, Russian Federation

E-mail: AOTolokonskij@mephi.ru

Abstract. The experience of the creation tools to support the construction of control systems design skills to provide effective training for students of the Department of Automation MEPhI, to create management information systems based on PCS. Currently the control system have been widely used not only in industry, but also on research. Therefore, quite important question of training in research automation.

1. Introduction

Developed and implemented in the educational process of laboratory practical work on the design of dynamic objects management systems. Practice allows you to fully simulate the system design process and includes the basic hardware and software modules necessary for the implementation of the design process. Support Tools were used in the creation and implementation of real control systems on a number of objects is not only the nuclear industry, but also in other areas of production. Created a set of tools allows you to create individual tasks for students to conduct innovative workshop. [1]

2. Experimental setup

The main part laboratory practical (is presented in figure 1):

- computer network that includes 10 or more computers - automated workplaces (AWP) stands for learners;
- one or two computers - ARM teachers, designed to monitor the implementation of laboratory work, including receiving data from the ARM stands;
- universal simulator facilities management;
- measuring devices - thermocouple, RTD and normalized sensors;
- software for generating jobs and control laws (separately on each computer);
- software framing video;
- software programming process, focused on their use of technology.

Content from this work may be used under the terms of the Creative Commons Attribution 3.0 licence. Any further distribution of this work must maintain attribution to the author(s) and the title of the work, journal citation and DOI.

Published under licence by IOP Publishing Ltd

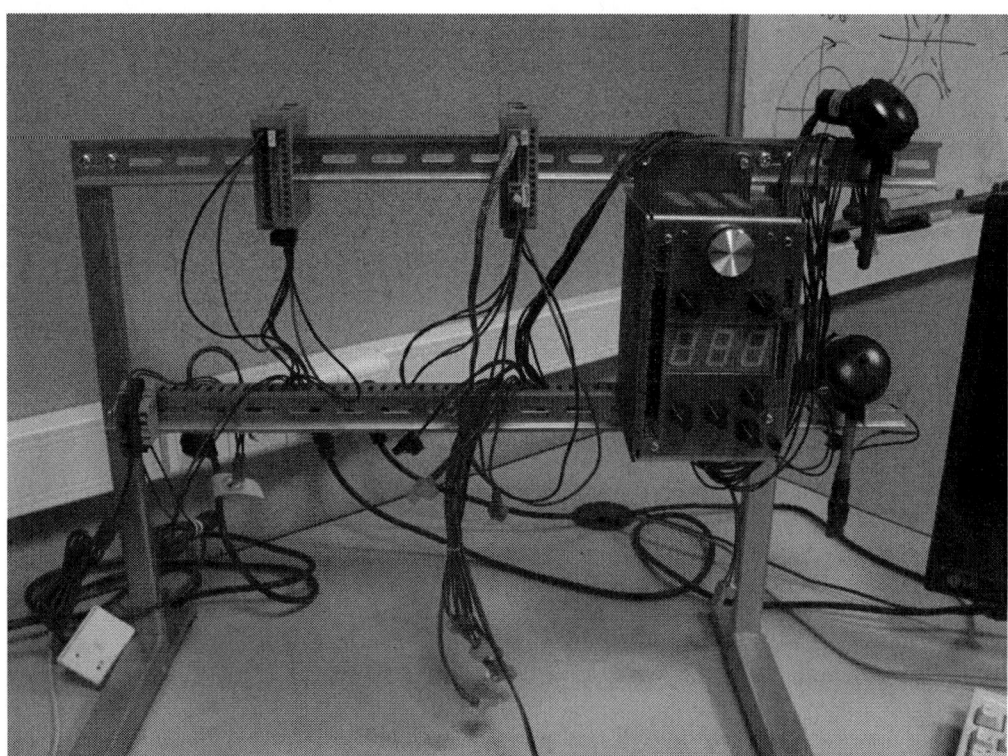

Figure 1. The complex technical equipment mini APCS.

Laboratory booth includes:

- converter RS485-USB;
- freely programmable CPU module PC100;
- monoblock MB100 (contains a complete set of analog and digital inputs and outputs, various functions PID control and refinement of locks);
- thermocouple;
- a heating element;
- universal power simulation facilities management;
- personal computer mounted with a PTC "UMIKON";
- connecting wires;
- a teaching aid for the implementation cycle of laboratory works with a detailed description of the complex software and hardware.

Object management (is shown in figure 2) consists of two tanks with a variable volume. In the first tank - mixing tank - serves two pipelines with hot and cold water. The costs of water supplied to the mixing tank is changed by means of the flap and the valve. The flap has a digital control (open / close), and can be controlled by turning the handle and is integrating the actuator type. The valve is controlled by the analog signal and is proportional to the actuator. [2] To address the various situations in the technology management model it is possible to change the inertia of the drain water temperature sensor, as well as the change in length of the pipeline draining, the master transport delay. Presented control object is implemented as emulation block. [3]

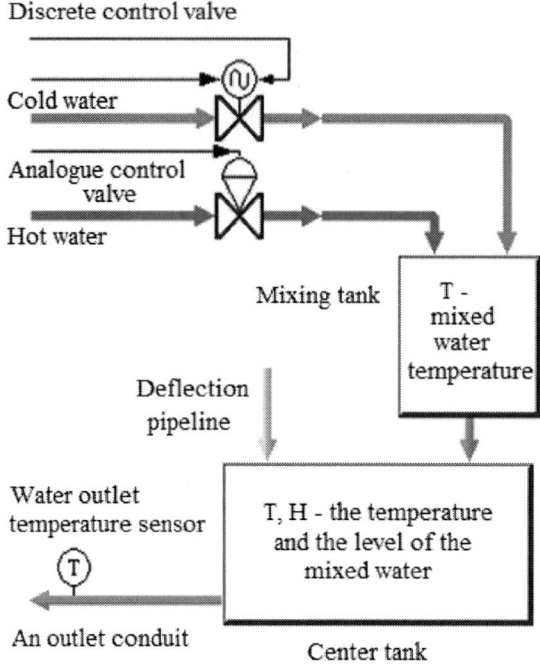

Figure 2. Block diagram of the control object model.

With block emulation control objects can be learned to build and configure almost all types encountered in the production control laws. [4] The output pattern output by analog watch costs mixing tank through the filling valve and the valve, the level in the mixer, the water temperature in the main tank and the temperature sensor readings at the end of the drain pipe from the main tank, the water flow out of the pipe "perturbation". Simultaneously, the built-in object model display control panel to monitor the level in the main tank, the temperature at the outlet of the mixer tank and the main tank. With block emulation object management study can be implemented, the construction and setting up of virtually all types encountered in the manufacture of regulators:

- the liquid level in the tank: the object of the first and second orders, integrating with the actuator and the proportional type, cascade controller;
- temperature: the object of the first and second orders, for object transport delay, to outline the perturbation;
- multivariable control - maintain the desired water temperature and the set level at the same time;
- tracking systems or dosing - maintaining a flow rate in proportion to the other.

Each student must carry out appropriate studies and calculations in accordance with the schedule and the individual job plan. Based on these studies Annotation reports should be compiled. Reports should include the theoretical calculations, the results of the identification of objects of management, the results of the control system for the archive data and the conclusions of the work done. The annexes to the instructions given guidance on theoretical calculations and practical work, as well as a list of recommended literature for self-study. There are 25 variants of tasks. Quest is designed for two semesters. In the first semester, students will have obtained on the basis of the job do the following studies:

- Preparation of job design in accordance with the individual tasks. Implementation of the control system on the PTC. Creating a mathematical model of the control object.

- Carrying out the identification of the control object on the basis of the model. Theoretical calculation of the stability test of the control object. Preparation annotation report for the certification of the first section of the discipline.
- Selection of the correction device. Theoretical calculation of the parameters of the correction device.
- Check the stability of the control system, the determination of the safety factor. Preparation annotation report for the certification of the second section of disciplines.
- In the second semester, you should:
- How to configure a control system to verify the theoretical calculations of the control parameters of the system.
- Operation of sales management system. Checking the system behavior during the transition from set point to set point. And the experiments to simulate external disturbances in the control system. Preparation annotation report.
- Further development of results-based management system operation. The practical realization of human-machine interface control system. Preparation of final report.

To successfully configure the control parameters necessary to make the students to identify the control object. This is necessary to make an experiment on registration of accelerating characteristics of the control object. [5] Acceleration (acceleration ramp) is a graph of the controlled variable changes over time as a result of an abrupt perturbation applied to the object. The greatest practical interest is the study of the dynamic properties of the control object under perturbations of the regulatory impact. To record the acceleration curve object is brought into an equilibrium state in which all of the input and output values are constant. Then a quick movement of the Regulator gives an abrupt perturbation. After that periodically record the results of measuring the output value as long as the output value will not accept a new steady-state value or install a constant rate of change. As reference points build the curve in the coordinates: the output value - time, which is the acceleration curve. On the basis of the transient response using one of the methods of identification, the students get the dynamic characteristics of the control object. And then are the parameters for the implementation of the model control law. This workshop provides simulation analysis capabilities for the correctness of the data of different identification methods. The results of the identification cannot always be close to the correct description of the object.

3. Discussion
Initial experience in conducting studies on the creation of management systems has shown that students pass a final training course, listening to and successfully passed specialization subjects may not always use the theoretical knowledge for the successful implementation of the task.

Acknowledgments
This work was supported by Competitiveness Program of National Research Nuclear University MEPhI.

References
[1] Lebedev V O, Komisarchuk S Y, Obnosov A V 2004 The structure and main features of the software package creation "MikSIS" DCS control systems "UMIKON" *Industrial ASU and controllers* №1 pp 35-41 (in Russian)
[2] Rao C R 1968 *Linear Statistical Methods and its Applications* (Moscow: Science) (in Russian)
[3] Vlasov V A, Vlasova S V, Tolokonsky A O 2014 Estimates when parametrical hypotheses are accepted and comparison them with maximum likelihood estimates *Life Science Journal* №11(11s)
[4] Seber G A F 1977. *Linear Regression Analysis, Wiley Series in Probability and Mathematical Statistics* (New York-London-Sydney-Toronto) p 456
[5] Vlasov V A and Vlasova S V 2007 Software Support for Testing Composite Hypotheses about the Parameters of Exponential Distribution taking into Account the Weighs of Errors *Instruments and Systems: Monitoring, Control, and Diagnostics* №8 pp 25-7 (in Russian)

Polymer-Free Carbon Nanotubes Saturable Absorbers for Nanosecond Pulse Generation

A V Zasedatelev[1,2], V I Krasovskii[2,4], O Reynaud[3], Yu G Gladush[1], D S Kopylova[1], E A Komochkina[4], E I Kauppinen[3] and A G Nasibulin[1,3,5]

[1]Skolkovo Institute of Science and Technology, Moscow, Russia
[2]Prokhorov General Physics Institute of the Russian Academy of Sciences, Moscow, Russia
[3]Department of Applied Physics, Aalto University School of Science, Espoo, Finland
[4]National Research Nuclear University «MEPhI», Moscow, Russia
[5]St.Petersburg State Polytechnical University, St. Petersburg, Russia

E-mail: anton.zasedatelev@gmail.com

Abstract. Hereby we present the results of investigations of nonlinear optical properties of single-walled carbon nanotube (CNT) thin-film saturable absorbers without binding polymers. Developed CNT-based polymer-free saturable absorbers exhibit high third-order nonlinear susceptibility: $\chi^{(3)} \sim 10^{-8}$esu, low absorption saturation intensity: $I_S \sim 30$ mW/cm^2, and high photostability. Using CNT-based polymer-free saturable absorbers for passive Q-switching mode of Nd:YAG laser, 25 ns laser pulses have been obtained.

1. Introduction

Carbon nanostructures are potentially promising nonlinear optical materials in laser physics. Due to a significant absorption saturation effect, graphene and CNTs are widely used as passive saturable absorbers for generation of ultrashort laser pulses both in fiber and solid-state laser systems [1,2]. Extremely high relaxation rate of the nonlinear response and large modulation depth of the absorption along with simple, flexible and low cost production of CNT-based materials are the main reasons of their rapid development in laser physics. Moreover, CNTs can be directly deposited on the optical fiber end or mirror, allowing their implementation in both classical and ring linear resonators [3]. Due to their high non-saturable losses, CNT-based saturable absorbers (CNT SA) become widely used in fiber lasers rather than in solid state systems. The elements, typically formed by deposition of CNT incorporated polymer suspension directly on the fiber end or quartz substrate, have several drawbacks. The dominant one is irreversible degradation of the polymer host of CNT SA, which caused by high intensity of laser radiation in the cavity. In order to prevent degradation processes in CNT SA polymer-free preparation methods have been developed [4].

The paper examines saturable absorption effect in polymer-free thin films of CNTs. Nonlinear optical measurements were carried out using open-aperture Z-scan technique, where 10 ns Nd:YAG laser (1064 nm) was used as a light source. We found out two mechanisms of nonlinear absorption, which related to fraction of CNT showing semiconductor and metallic properties. Nonlinearity of semiconductor CNTs can be described in two-level approximation, whereas nonlinearity of metallic CNTs is well described using third-order nonlinear susceptibility (Kerr nonlinearity). Here we

Content from this work may be used under the terms of the Creative Commons Attribution 3.0 licence. Any further distribution of this work must maintain attribution to the author(s) and the title of the work, journal citation and DOI.
Published under licence by IOP Publishing Ltd

estimated saturation intensity and nonlinear susceptibility through the numerical fitting of Z-scan data. As a result of high nonlinear optical properties and photostability of polymer-free CNT SA, 25 ns laser generation in Nd:YAG laser was obtained.

2. Formation of thin-film saturable absorbers

The CNTs were synthesized by an aerosol (floating catalyst) CVD method by two various approaches. The first one is based on thermal vapor decomposition of ferrocene in carbon monoxide atmosphere at the temperature of 875 °C [4]. Hereinafter, the product, single-walled CNTs produced by this method, will be named as Sample 1. In the second approach the feedstock solution contained ferrocene as a catalyst particle precursor, toluene and ethylene as carbon sources and thiophene as a promoter was introduced in the reactor in a hydrogen atmosphere and heated up in the reactor. The CNTs synthesized at 1100 °C with a 17 sccm ethylene flow and a molar ratio of sulfur and iron of 1:1 (ferrocene and thiophene concentrations of 0.5 % wt.) were a mixture of single-walled and double-walled CNTs (Sample 2) [5]. In both cases the product was collected downstream of the reactor by filtering the flow in the form of thin films with adjustable thicknesses (transmittance) and subsequently transferred on a desirable substrate by a dry transfer technique [6]. A piece of a nitrocellulose membrane filter with a CNT-film was placed on a quartz substrate with the CNT-film upside down. Then, the substrate and the filter were pressed together at the pressure of 1000 Pa. After the pressing procedure, the membrane filter was peeled off and the CNT- film was strongly adhered to the mirror surface. It is worth noting that the CNT-film was utilized as-deposited and no purification or dispersion steps were required. When compared to standard wet deposition methods, which may require several time-consuming stages, such as purification, dispersion and filtering, the approach of the CNT-film preparation demonstrated here is simple and easily scalable [7].

Following the abovementioned procedure two types of CNT thin films have been prepared. Sample 1 exhibits resonant absorption at 1050 nm, whereas Sample 2 has minimum in absorption at this spectral region (figure 1A and 1B respectively). Thus, the irradiation of Nd:YAG laser (λ=1064 nm) is reputed to be resonant and nonresonant with regard to absorption of Sample 1 and 2, respectively. The absorption spectra of both samples are shown in figure 1.

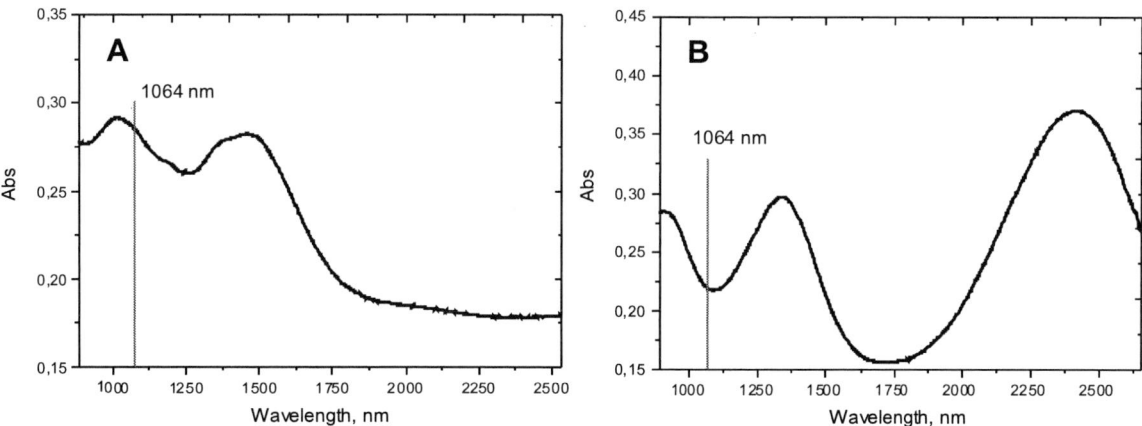

Figure 1. Absorbance spectra of Sample 1 – A and Sample 2 – B.

Absorption maxima of the samples are the result of interband (in case of semiconductor CNTs) and intraband (in case of metallic CNTs) transitions.

3. Nonlinear optical properties

Experimental studying of nonlinear optical absorption was carried out using open-aperture single beam Z-scan technique [8]. We utilised the first harmonic generation light of mode-locked Nd:YAG laser (1064 nm) operated in TEM$_{00}$ mode with the pulse duration of 10 ns at a low repetition rate of 4 Hz (to prevent heating processes). The beam was tightly focused by a lens, the beam waist was 30 μm with the pulse energy 85 μJ. The peak intensity in the focus (I_0) was 340 MW/cm^2. Z-scan curves are shown in figure 2.

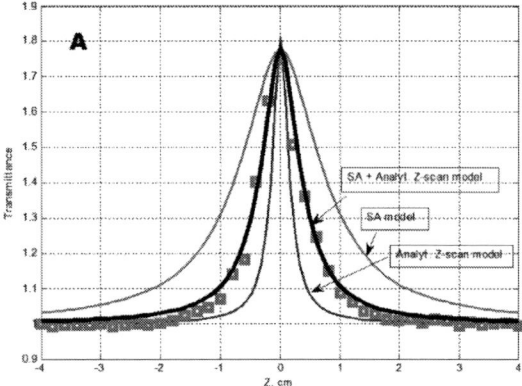
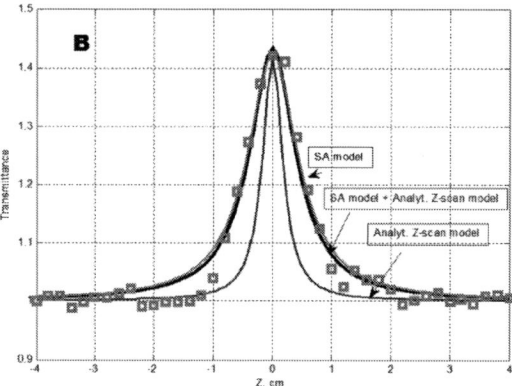

Figure 2. Open-aperture Z-scan curves of Sample 1 (A) and Sample 2 (B). Experimental data – red dots, model curves based on two-level saturable absorber approach (SA model) – red lines, model curves based on common used Sheik-Bahae formalism (Analyt. Z-scan model.) – blue lines, model curves based on both approaches (SA + Analyt. Z-scan model) – black lines.

As can be seen in figure 2, both samples exhibit considerable absorption saturation effect. In order to interpret experimental results we applied several fitting approaches. Well-known analytic model based on Sheik-Bahae formalism [8], numerical model of saturable absorber based on two-level approximation [9] and joint numerical model based on both approaches were considered [10]. The latter one is appropriate for nonlinear media having complicated mechanisms of the nonlinearity, such as ensembles of CNTs, and gives an opportunity to determine saturation intensity and third-order nonlinear susceptibility as well. A number of studies have shown the effectiveness of this approach to determine the nonlinear optical properties of SWCNTs and graphene by fitting Z-Scan curves [11,12].

Analytic Z-scan model (Analyt. Z-scan model) is derived from the solution of nonlinear wave equation in the paraxial approximation, where the beam profile is assumed to be Gaussian [8]. Therefore, the model is not appropriate for fitting Z-scans of thick samples and samples with a high phase nonlinearity. The resultant normalized transmittance for a pulse with a Gaussian profile can be written as:

$$T(Z) = \sum_{m=0}^{\infty} \frac{(-\beta_{eff} I_{00} L_{eff})^m}{(1 + (\frac{Z}{Z_0})^2)^m (m+1)^{3/2}} \qquad (1)$$

where $I_{00} = I(Z = 0, r = 0, t = 0) = \dfrac{4E_p}{3\pi \cdot \tau_p \cdot \varpi_0^2}$ is the on-axis intensity at focus ($Z=0$) at $r=0$, $t=0$,

E_p is the pulse energy, τ_p is the pulse duration, ϖ_0 is the beam waist, β_{eff} is the effective third-order nonlinear absorption coefficient, $L_{eff} = (1 - e^{-\alpha_0 L})/\alpha_0$ is the effective sample thickness, α_0 is the linear absorption coefficient, L is the actual sample thickness, $Z_0 = \dfrac{\pi \varpi_0^2}{\lambda}$ is the Rayleigh length, λ is the laser wavelength.

According to the two-level saturable absorber model (SA model), the absorption coefficient is determined by the following expression:

$$\alpha = \frac{\alpha_0}{1 + I/I_S} \tag{2}$$

where I_s is the saturation intensity, α_0 is the linear absorption coefficient.

The output intensity is determined by numerical solution of the following differential equation:

$$\frac{dI}{dL} = I \cdot \exp[-\alpha \cdot L] \tag{3}$$

Normalized transmittance can be obtained by numerical spatial and temporal integration of the input and output intensities:

$$T(z) = \frac{\int\limits_{-\infty}^{\infty} dt \int\limits_{0}^{\infty} I_{Out} r dr}{\int\limits_{-\infty}^{\infty} dt \int\limits_{0}^{\infty} I_{In} r dr} \tag{4}$$

where $I_{Out}(Z, r, t)$ is output intensity (which is determined from equation 3), $I_{In}(Z, r, t)$ is the input intensity.

The joint model (SA + Analyt. Z-scan model) involves properties of both two-level saturable absorber, and a Kerr-like nonlinear medium. According to the model, the absorption coefficient consists of slow term related to two-level saturable absorber and fast third-order nonlinear contributions:

$$\alpha = \frac{\alpha_0}{1 + I/I_S} + \beta_{eff} \cdot I \tag{5}$$

In that case normalized transmission is also determined by numerical integration of the intensities (in accordance with equation 4), where the output intensity can be obtained by solving the differential equation 3.

Fitting was performed with least-square method where the nonlinear absorption coefficient β_{eff} and the saturation intensity I_S were considered as variable parameters.

The fitting results are presented in figure 2. As can be seen, the joined model provides the best coincidence with experimental data. Especially it gives the best fit for Z-scan of Sample 1 (figure 2A), the analytic and the two-level saturable absorber model are not able to fit the data. We suppose that Sample 1 consists of both types of CNTs exhibiting semiconductor and metallic properties, which provide slow and fast nonlinearity respectively. Slow nonlinearity is associated with the saturation of interband transition in the semiconductor fraction, meanwhile the fast is related to excitation of hot electrons in the metallic fraction. Since the Z-scan data of Sample 2 can be well fitted by two-level saturable absorber model, the fraction of CNTs with its metallic properties is negligible, i.e. nonlinear response in that case is attributed only to semiconductor CNTs. The best fit nonlinear optical parameters for both samples are listed in table 1.

Table 1. Nonlinear optical properties of polymer-free CNT SA

	I_s [mW/cm^2]	β_{eff} [m/W]	$\chi^{(3)}$ [esu]
Sample 1	30	- 6·10^7	- 1,3·10^{-8}
Sample 2	35	- 1,2·10^7	- 3,4·10^{-9}

In accordance with the obtained nonlinear optical parameters, the saturation intensity is nearly the same for both samples, but imaginary part of third-order susceptibility is about 5-times larger in case of Sample 1. It means a higher nonlinearity of Sample 1 at high input intensities. In fact, the fraction of CNTs having metallic properties provides larger modulation depth ΔT of Sample 1 (76%) in comparison with the depth of Sample 2 (42%) at the focus (intensity at the focus equals 340 MW/cm^2). The modulation depth at lower intensities (33 MW/cm^2) is nearly the same for both samples: 14.5% and 11.5% for Sample 1 and Sample 2 respectively. Therefore, we can conclude that both samples with resonant and nonresonant absorption are equally applicable for utilizing in low- and medium-power laser systems, whereas the Sample 1 exhibiting resonant optical absorption is more appropriate for high power systems than Sample 2.

4. Generation of nanosecond pulses

In order to examine a potential of using polymer-free CNT SA for passive Q-switching in solid state lasers with high pulse energy, we built up the following experimental setup based on Nd:YAG laser (figure3).

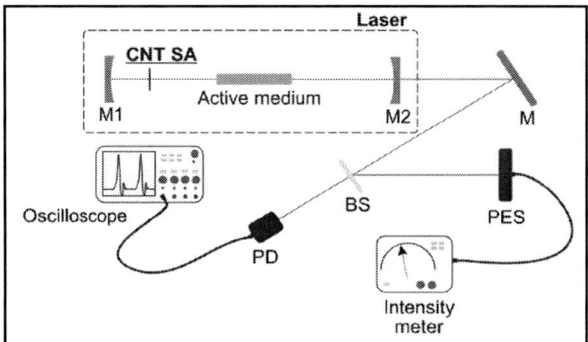

Figure 3. The scheme of experimental setup for laser generation with usage of polymer-free CNT SA.

Silicon photodetector with integrated signal amplifier (HFBR-25X6Z Series Avago Technologies) with a rise time of 3.3 ns, Rigol DS12048 oscilloscope with a bandwidth of 200 MHz and Ophir PE25-C pyroelectric energy sensor (PES) within the Pulsar power meter interface were used to measure the laser output performances. We employed 35 cm linear cavity and flash-pumped Nd:YAG active medium. Without CNT SA laser was operated in free-running mode (figure 4A), where an average duration of the overall pulse train and individual pulse spikes are equal to 100 μs and 600 ns respectively.

Considerable pulse narrowing effect and enhancement of the pulse intensity were obtained when Sample 1 was placed into the cavity (figure 4B). It clearly indicates the transition in the regime of laser generation: from free-running to Q-switching mode.

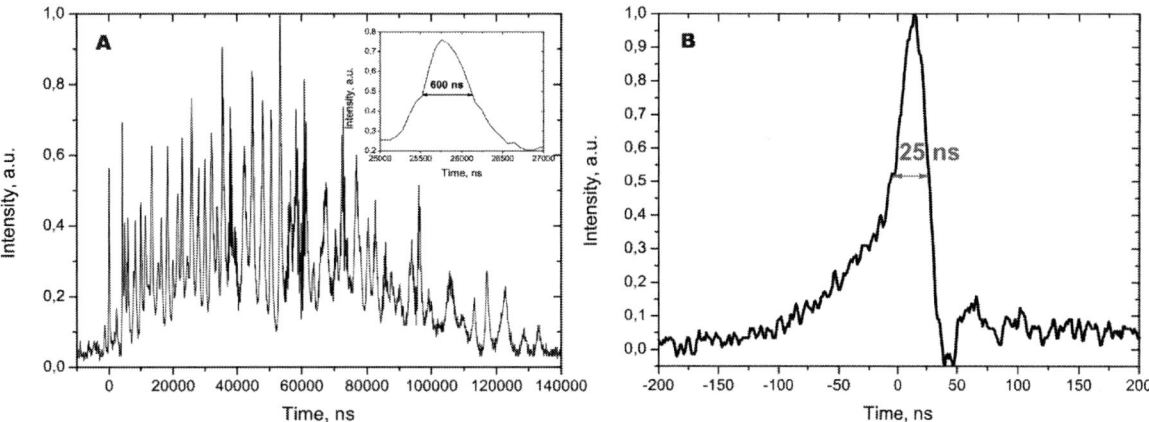

Figure 4. Oscilloscope trace of laser output in free-running mode (without CNT SA) – A. Inset: Oscilloscope trace of single pulse spike in free-running mode. Oscilloscope trace of laser output in Q-switching mode with usage of polymer-free CNT SA – B.

5. Conclusion

We have studied nonlinear optical absorption of polymer-free CNT SA by Z-scan technique using nanosecond 1.06 μm laser irradiation. Complex nonlinear response of the CNT SA was found. Thus, at low input intensities the response can be well described by ordinary saturable absorber model with saturation intensity equal to 30 MW/cm², but at the high intensity the response is described by third-order Kerr nonlinear susceptibility $\chi^{(3)} \sim 10^{-8}$ esu. We suppose that such behavior is induced by presence of two types of CNTs in the samples with semiconductor and metallic properties. Developed polymer-free CNT SA demonstrate large saturable absorption effect and high photostability, while retaining their properties during long-term exposure of nanosecond pulses with an average energy ~ 100 μJ. In summary, using polymer-free CNT SA we successfully implemented passive Q-switching mode of solid state Nd:YAG laser with a pulse duration of 25 ns.

Acknowledgements

This work has been supported by the Ministry of Education and Science of Russian Federation (Project No. RFMEFI58114X0006).

References

[1] Martinez A and Sun Z 2013 Nanotube and graphene saturable absorbers for fibre lasers *Nature Photon.* **7** 842

[2] Cho W B, Yim J H, Choi S Y, Lee S, Schmidt A, Steinmeyer G, Griebner U, Petrov V, Yeom D-I, Kim K and Rotermund F 2010 Carbon nanotubes: boosting the non linear optical response of carbon nanotube saturable absorbers for broadband mode-locking of bulk lasers *Adv. Funct. Mater.* **20** 1937

[3] Set S Y, Yaguchi H, Tanaka Y and Jablonski M 2004 Laser mode locking using a saturable absorber incorporating carbon nanotubes *J. Lightwave Technol.* **22** 51

[4] Moisala A, Nasibulin A G, Brown D P, Jiang H, Khriachtchev L and Kauppine E I 2006 Single-walled carbon nanotube synthesis using ferrocene and iron pentacarbonyl in a laminar flow reactor *Chem. Eng. Sci.* **61** 4393

[5] Reynaud O, Nasibulin A G, Anisimov A S, Anoshkin I V, Jiang H and Kauppinen E I 2014 Aerosol feeding of catalyst precursor for CNT synthesis and highly conductive and transparent film fabrication *Chem. Eng. J.* **255** 134

[6] Kaskela A *et al* 2010 Aerosol synthesized SWCNT networks with tuneable conductivity and transparency by dry transfer technique *Nano Lett.* **10** 4349

[7] Zhou Y, Hu L and Grüner G 2006 A method of printing carbon nanotube thin films *Appl. Phys. Lett.* **88** 123109

[8] Sheik-Bahae M, Said A and Van Stryland E W 1990 Sensitive measurement of optical nonlinearities using a single beam *IEEE J. Quantum. Electron* **26** 760

[9] Wang F, Rozhin A G, Scardaci V, Sun Z, Hennrich F, White H, Milne W I and Ferrari A C 2008 Wideband-tuneable, nanotube mode-locked, fibre laser *Nat. Nanotechnol.* **3** 738

[10] Karthikeyan B, Anija M and Philip R 2006 In situ synthesis and nonlinear optical properties of Au:Ag nanocomposite polymer films *Appl. Phys. Lett.* **88** 053104

[11] Kamaraju N, Kumar S, Sooda A K, Guha S, Krishnamurthy S and Rao C N R 2007 Large nonlinear absorption and refraction coefficients of carbon nanotubes estimated from femtosecond z-scan measurements *Appl. Phys. Lett.* **91** 251103

[12] Liu Z, Wang Y, Zhang X, Xu Y, Chen Y and Tian J 2009 Nonlinear optical properties of graphene oxide in nanosecond and picosecond regimes *Appl. Phys. Lett.* **94** 021902

Institute of Physics
Dirac House, Temple Back
Bristol BS1 6BE UK

ISSN: 1742-6588
ISBN 978-1-5108-2915-2

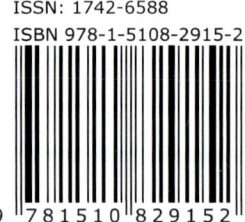

9 781510 829152